Lecture Notes in Geosystems Mathematics and Computing

Series Editors
W. Freeden, Kaiserslautern, Germany
M. Z. Nashed, Orlando, FL, USA
O. Scherzer, Vienna, Austria

 Birkhäuser

More information about this series at http://www.springer.com/series/15481

Willi Freeden • Clemens Heine • M. Zuhair Nashed

An Invitation
to Geomathematics

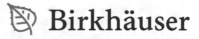 Birkhäuser

Willi Freeden
Mathematics Department
University of Kaiserslautern
Kaiserslautern, Germany

Clemens Heine
Executive Editor
Springer Nature
Heidelberg, Germany

M. Zuhair Nashed
Mathematics Department
University of Central Florida
Orlando, USA

Lecture Notes in Geosystems Mathematics and Computing
ISBN 978-3-030-13053-4   ISBN 978-3-030-13054-1   (eBook)
https://doi.org/10.1007/978-3-030-13054-1

Library of Congress Control Number: 2019934803

Mathematics Subject Classification (2010): 01A05, 31C05, 35R30, 35Q86, 45Q05, 76U05

This book is published under the imprint Birkhäuser, www.birkhauser-science.com by the registered company Springer Nature Switzerland AG.
The registered company address is: Gewerbestrasse 11, 6330 Cham, Switzerland

Preface

In spring 2008, on the occasion of Willi Freeden's 60th birthday, Volker Michel organized a symposium on "Geomathematics" in the Rotunde of the University of Kaiserslautern with about 50 participants from all over the world. After 2 days lecturing by distinguished people from all geoscientific disciplines, M. Zuhair Nashed made the following remarks during the conference dinner:

> Willi, the conference is fantastic, the lectures of our colleagues show the potential and beauty of geomathematics, however, it is amazing that there is no international journal on geomathematics with a high reputation, although there are so many outstanding researchers. Probably, Springer is a good address to run such a journal!

Motivated by these comments, Willi contacted Springer and started a dialogue on this issue with Clemens Heine from Springer's Mathematics Editorial. Clemens proposed to discuss the pros and cons for such a new journal on geomathematics after the "DMV Jahrestagung, Erlangen, 2008," during a train ride from Nürnberg to Mannheim. The result of the very intense "train discussion" was the agreement to examine the market for geomathematical obligations by first introducing a *Handbook of Geomathematics*. Subsequently, Willi convinced M. Zuhair Nashed and Thomas Sonar to act as joint editors, and the handbook indeed became a valuable work in the geoscientific community. The first edition of the handbook published in 2010 contained 45 contributions, the second edition published in 2015 showed even the double number of papers.

Based on the success of the *Handbook of Geomathematics* and the positive feedback of the mathematical community, the first Springer issue of the *GEM International Journal on Geomathematics* was published in August 2010 with Willi as Editor-in-Chief and an elected editorial board of scientists across the globe. Today, after more than 10 years of geomathematical research involving *GEM International Journal on Geomathematics* as a forum for high-level publication, the stage is set for a résumé on the area of geomathematics itself.

Accordingly, this book can be seen as a survey on the broad landscape of geomathematical research, its major topics and challenges, as well as a resulting manifestation what geomathematics had, has, and will have to offer. Both mathematicians and geoscientists with mathematical affinity are invited to consider and use the existing geomathematical facilities as an adequate platform for successful research and sustainable developments for the future.

Kaiserslautern, Germany   Willi Freeden
Heidelberg, Germany   Clemens Heine

Willi Freeden and Geomathematics

The term "geomathematics" seems to have been introduced by Willi Freeden to the literature in his joint book with T. Gervens and M. Schreiner entitled *Constructive Approximation on the Sphere (with Applications to Geomathematics)*, Oxford Science Publication, Clarendon, 1998. One can surmise from reading this book that Willi meant the following:

Geomathematics is the latest newcomer to the arena of Geo-X. First, we had "geophysics" which deals with physics of the Earth. Then, we had "geostatics" which was mainly driven by applications of statistics to mining. "Geostatistics", broadly defined, deals with stochastic developments and statistical validation to describe Earth's sciences phenomena. Then, we had "Geoinformatics" which deals with information in a georeflected environment. And now we have "geomathematics".

Willi Freeden founded the Geomathematics Group at the University of Kaiserslautern in 1994. He identified the kernel of geomathematics and played a major role in developing its landscape over the last 25 years. This was achieved by his research and the work with his students and collaborators. He supervised 36 PhD dissertations in the field of geomathematics (listed in Appendix D) and many diploma and master's theses. The research conducted at the Institute for Geomathematics was disseminated by the widely circulated technical reports. Many of the dissertations of his students were published as books. Willi Freeden published 19 books (see Appendix C). *GEM: International Journal on Geomathematics* was founded by Willi Freeden, who has spent a lot of time and effort as Editor-in-Chief to establish it as a high-quality journal.

Several international conferences have been organized by Willi Freeden at Oberwolfach, Kaiserslautern, and other locations, and by Volker Michel—celebrating anniversaries of Willi.

All these activities have established "geomathematics" as the bridge between mathematics and the geosciences. Willi has played a pivotal role in all of these activities.

On a personal note, I have been fortunate to know Willi and to collaborate with him in research and to organize several minisymposia and special sessions at meetings of the American Mathematical Society. His generosity and innovative ideas have been a great inspiration for me. It is a great honor for me to collaborate with Willi.

Orlando, USA   M. Zuhair Nashed

Abstract

During the last few decades, all geosciences were influenced by two essential scenarios: First, the technological progress has completely changed the observational and measurement techniques. Modern high-speed computers and satellite-based techniques are entering more and more disciplines. Second, there was a growing public concern about the future of our planet, i.e., the change of its climate, the obligations of its environment, and about an expected shortage of its natural resources. Simultaneously, all these aspects implied the strong need of adequate mathematical structures, tools, and methods, i.e., *geomathematics*.

The goal of this contribution is to characterize today's geomathematics as key technology with respect to methodological origin and geoscientific foundation, constituting ingredients, scientific core, as well as perspective potential. Novel exemplary applications are proposed in the field of geoexploration.

The role of *GEM International Journal on Geomathematics* as forum and organ for geomathematical research is highlighted in an Appendix A. Moreover, a list of the contributions of the second edition of the *Handbook of Geomathematics* is given in Appendix B.

Keywords Mutual interplay of geosciences and geomathematics, Geomathematics: its role, its potential, and its perspective, Exemplary applications in geoexploration, GEM International Journal on Geomathematics as geoscientifically relevant mathematical forum and organ, Handbook of Geomathematics

Abstrakt

Während der letzten Dekaden waren die Geowissenschaften von zwei wesentlichen Szenarien beeinflusst: Zum einen hat der technologische Fortschritt die Beobachtungs- und Messmethoden vollständig geändert. Moderne Hochleistungsrechner und satellitenbasierte Techniken kamen mehr und mehr zum Zuge. Zum anderen gab es eine wachsende Besorgnis in der Bevölkerung um die Zukunft unseres Planeten, d.h. den Wandel des Klimas, die Belange seiner Umwelt und die erwartete Verknappung seiner natürlicher Ressourcen, Alle diese Aspekte implizieren simultan den starken Bedarf an adäquaten mathematischen Strukturen, Hilfsmitteln und Methoden, kurzum an *Geomathematik*.

Der vorliegende Beitrag widmet sich der Aufgabe, die heutige Geomathematik als Schlüsseltechnologie hinsichtlich methodologischem Ursprung und geowissenschaftlicher Grundlegung, konstituierender Bestandteile, wissenschaftlicher Rolle sowie perspektivischem Potential zu charakterisieren. Neuartige beispielhafte Anwendungen werden im Bereich der Geoexploration vorgeschlagen.

Die Rolle der Zeitschrift *GEM International Journal on Geomathematics* als Forum und Organ für geomathematische Forschung wird im Appendix A beleuchtet. Darüber hinaus liefert Appendix B alle Beitragstitel der zweiten Edition des *"Handbook of Geomathematics."*

Layout of the Book

This book intends to introduce the science of geomathematics to a wider community and to promote the *GEM International Journal on Geomathematics*.

- Chapter 1 presents an introduction to the Earth as a system to apply scientific methods and structures. Emphasis is laid on transfers from virtuality to reality, and vice versa.
- In Chap. 2, geomathematics is introduced as a new scientific area which nevertheless has its roots in antiquity. The modern conception of geomathematics is outlined from different points of view, and its challenging nature is described as well as its interdisciplinarity. Geomathematics is shown as the bridge between the real world and the virtual world. The complex mathematical tools are shown from a variety of fields necessary to tackle geoscientific problems in the mathematical language.
- Chapter 3 contains some exemplary applications as novel exploration methods. Particular importance is laid on the change of language when it comes to translate measurements and observations to mathematical models. New solution methods like the multiscale mollifier technique are presented for the purpose of geoexploration, such as inverse gravimetry and reflection seismics.
- Chapter 4 is devoted to the short description of recent activities in geomathematics, e.g., the *S*ociety of *I*ndustrial and *A*pplied *M*athematics (SIAM) program on "Geosciences and Mathematics of Planet Earth."
- The Appendix A is devoted to the *GEM International Journal on Geomathematics* founded ten years ago. Besides a detailed structural analysis of the editorial goals, an index of all papers published in former issues is given.
- The Appendix B gives a list of all titles of the contributions published in the "Handbook of Geomathematics" (2015).
- The Appendix C delivers a collection of published books and edited handbooks by Willi Freeden.
- The Appendix D presents a list of geomathematically relevant PhD-theses supervised by Willi Freeden.

In order to explain the role, aims, and potential of geomathematics within Earth system research, we frequently follow the work given, e.g., in [26–28]. We also use some explanations and illustrations from the article [47] given in the *Handbook of Geodesy* (Special Issue: "Mathematical Geodesy").

Contents

Chapter 1
Earth's Science System

Humankind today is changing essential components of the Earth's system including the climate system, without sufficient understanding of the constituting ingredients. There is an urgent need for a better knowledge of the Earth's system (Fig. 1.1) and the interrelations among its components. This need cannot be satisfied without two requirements, *measurements and observations* of a large set of influencing parameters and *concepts and models* characterizing the Earth's science system.

A constituting element of Earth's system research is to explore and exploit the close connections between the process of geoscientific observation and the mathematical modeling of Earth's system constituents. The processes and phenomena of the Earth's system manifest themselves, e.g., in geophysical and geodetic parameters. Observations determine parameter sets and time series about the geometry and physics of the Earth. These parameters are incorporated into models and simulations of the Earth's system processes and thereby help to understand and improve the knowledge about the Earth's system. Conversely, models and simulations enable an improved mathematical analysis and a proper interpretation and application of measurements, which in turn provide innovations in measurement and observation techniques. There is a circuit of measurements and observations and geomathematical structures and concepts. Its overall aim is to obtain more accurate and more consistent results for Earth's system research.

The basis of Earth's system research are measurements and observations, i.e., scalar numbers, vectors, tensors such as distances, angles, directions, velocities, and accelerations. Nowadays, computer facilities and measurement and observation methods open new research areas and opportunities. However, it is a geoscientific trademark to present measured values always together with a suitable modeling procedure for interpretation and an appropriate knowledge and estimation about reliability and accuracy. So, inherently, mathematics is involved as key technology bridging the real world of measurements and the virtual world of handling datasets, modeling geoscientific quantities and processes, and providing illustrations and interpretations.

© The Author(s), under exclusive license to Springer Nature Switzerland AG 2019   1
W. Freeden et al., *An Invitation to Geomathematics*, Lecture Notes in Geosystems
Mathematics and Computing, https://doi.org/10.1007/978-3-030-13054-1_1

Fig. 1.1 The Earth as seen from the Apollo 17 mission in 1972 (cf. Apollo 17. https://web.archive.org/web/20160112123725)

The result of measurements are scalar numbers, vectors, and tensors, i.e., the raw material. Mathematical handling and approximation of datasets as well as modeling techniques are necessary to connect the "reality space" with the "virtuality space." In this sense, a model represents the result of the transfer, it intends to be an image of the reality, expressed in mathematical language, so that an interaction between abstraction and concretization is involved. The mathematical world of numbers and structures contains efficient tokens by which we are able to describe the rule-like aspect of a real problem. This description usually includes a simplification by abstraction: essential properties of, e.g., relevant problems are separated from unimportant ones and a solution scheme is set up. The "eye for similarities" enables mathematicians to recognize a posteriori that resulting solutions become applicable to multiple cases.

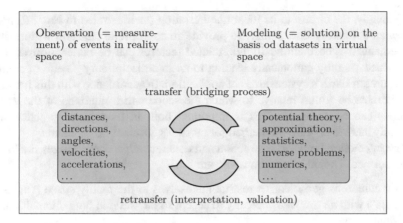

Fig. 1.2 The circuit "reality space (observation) and virtual space (modeling)"

Summarizing we are led to the following conclusion: *Earth's system research* is characterized by an interaction following a "circuit" (see Fig. 1.2):

- An input problem from *reality space* ("measurements") that, reduced by abstraction and transferred into *virtuality space*, results in a mathematical output model which becomes a new "concrete" input problem in *reality space*.

As a consequence, the ideal process (circuit) for the solution process of geoscientific problems canonically requires the following steps to be handled (see also the approach sketched in [27]):

- *Transfer from Reality to Virtuality Space:* Measurements and observational events in reality space lead to mathematical tokens and quantities as the "raw material" for modeling and processing in virtuality space. The observational input is translated into the language of the virtuality space, i.e., mathematics, requiring close cooperation between application-oriented and mathematical scientists.
- *Retransfer from Virtuality to Reality Space:* Appropriate analytical, algebraic, statistic, stochastic, and/or numerical methods must be taken into account; if necessary, new solution techniques must be proposed. The mathematical model is validated, the aim is a good accordance of model and measurement. If necessary, on the basis of new measurements, the model must be improved by use of modified "raw material" and/or by an alternative methodology.

Obviously, the benefit of such a circuit is a better, faster, cheaper, and more secure problem solution on the basis of the mentioned processes of modeling, simulation, visualization, as well as reduction, decorrelation, and denoising of large amounts of data. The more measurements are available, the more one recognizes the causality between abstraction by mathematical concepts and their impact and cross-sectional importance to reality.

Evidently, the circuit in its ideal manifestation (as illustrated in Fig. 1.2) has to follow an obligatory line, namely to provide an appropriate platform within which mathematically/geoscientifically interrelated features can be better motivated and understood, thereby canonically leading to an interdisciplinary palette of solution procedures in Earth's system areas of application. In accordance with this intention, criteria must be found relative to which the scope and limitations of the various methods can be assessed. This is important both in theory and practice since generally there is no cure-all method for most of geoscientific problems.

The interaction between abstraction and concretization characterizes the history of geoscience. The questions, however, are

- why numerous geoscientists restrict themselves to the reality space ("measurements") with an unloved necessity to accept some "service fundamentals" of the virtuality space,
- why numerous mathematicians are interested only in rare exceptions in appropriate handling of practically relevant geo-obligations including specific model developments.

Following an article about the interconnecting roles of geodesy and mathematics presented by Moritz [87], a prominent member of today's georesearch community, the actual interrelationship shows a twofold appraisal from history (see also [28]):

- First, Moritz [87] states that the old days are gone when Carl Friedrich Gauss (1777–1855) himself developed his epoch-making theories inspired by his geoscientific concerns. Gone also are the days when Felix Klein (1849–1925), one of the leading mathematicians of his time, called geodesy "that geometrical discipline in which the idea of approximation mathematics has found its clearest and most consequent expression" (see [71, p. 128]). Gone are the times when H. Poincaré (1854–1912) investigated problems of astronomy, geodesy as well as geophysics and actively participated in geoscientific life. So we are led to the conclusion that it apparently is the fault of today's mathematicians that they provide mathematics in an increasingly abstract way, without any regard to possible geoscientifically relevant applications and, so to say in the scheme of Fig. 1.2, out of touch with reality? Moritz' opinion is as follows: "In part, certainly, they are out of reality."
- Second, Moritz [87] is deeply convinced that an increasing abstraction is necessary to achieve progress, not only in mathematics, but also in today's geoscience. What is frequently overlooked, e.g., by potential geodetic users of mathematical theory is that the modern abstract methods of mathematics, if properly understood, provide an extremely powerful tool for the solution of applied problems which could not be solved otherwise: the more abstract a method is, the more it is sometimes suitable for a concrete problem. Thus, we may also conclude that apparently it is the fault of modern geoscientists to be restricted to measurement tasks, without any regard to virtuality space providing valuable mathematical concepts and, so to say, also out of touch with virtuality.

As a consequence, in the sense of Moritz' explications, today's circuits should follow the ideal way—at least to a considerable extent—that was initiated by C.F. Gauss as one of the history's most influential mathematicians and geoscientists for an extremely fruitful interdisciplinary exchange (cf. [52]). The heritage of Gauss's work has extremely much to offer even these days to build a strong scientific bridge between mathematics and geoscience by the consequent continuation of the interplay between abstraction and concretization.

Unfortunately, it must be confessed that today's circuits (in the sense depicted in Fig. 1.2) turn out to be too complex in their transfer demands from reality to virtuality space and vice versa, as to be handled by only one ingenious geo-scientist. In addition, Earth's system changes have been accelerated dramatically. Today, the appalling situation is that a large number of Earth's system problems in their specific changes and modifications over the last years must be solved simultaneously. Interdisciplinary solutions are urgently required as answer to an increasingly complex world. In the opinion of the authors, the scientific challenge is a "geoscientific consortium." Indeed, the leading role of mathematics for obligations in virtuality space must be acknowledged (again) within today's geoscience, so that mathematicians will become more enthusiastic about working on geoscientific programs. An "Earth's system consortium" reflecting the cross-sectional demands in reality as well as virtuality space is absolutely essential for a sustainable development in the future.

No doubt, mathematicians interested in Geosystems Mathematics (cf. Fig. 1.3) can and should be integrated smoothly into the geoscientific phalanx. Only an "Earth's system consortium" consisting of scientists with equal standing, rights,

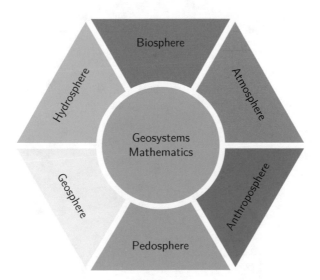

Fig. 1.3 Geosystems Mathematics as the key technology penetrating the complex system Earth (cf. [51])

and research position will be able to promote the significance of the Earth's system research in its responsibility even for society similarly to the Gaussian epoch.

All in all, Earth's system research today is confronted with a twofold situation:

- to make mathematicians aware of the particular measurement concepts and observational developments occurring in reality space,
- to make geoscientists conscious of new tools, means, structures, methods, and procedures for handling recent measurements and observations in virtuality space.

In fact, there is an urgent need for essential steps towards modern manifestations of "Earth's system consortia," realizing the cross-sectional demands and requirements of today's Earth system circuits in a well-balanced interdisciplinary way. In this respect, the characterization of the particular role of geomathematics is indispensible, and the descriptive imperative is one of the aims of this paper.

Chapter 2
Geomathematics

Mathematics (from Greek $\mu\acute{\alpha}\vartheta\eta\mu\alpha$ "knowledge, learning") intends to study topics as quantity, structure, space, and change. Correspondingly, $\gamma\epsilon\omega\mu\acute{\alpha}\vartheta\eta\mu\alpha$ (*geomathematics*) is $\mu\acute{\alpha}\vartheta\eta\mu\alpha$ (mathematics) concerned with geoscientific obligations. In our times, geomathematics is thought of being a very young science and a modern area in the realms of mathematics. However, nothing is further from the truth (cf. [113]). Geomathematics began as people realized that they walked across a sphere-like Earth and that this observation has to be taken into account in measurements and calculations. In consequence, we can only do justice to geomathematics, if we look at its historic importance, at least briefly (cf. [26, 27]).

According to the oldest evidence which has survived in written form, geomathematics was developed in Sumerian Babylon and ancient Egypt (see Fig. 2.1) on the basis of practical tasks concerning measuring, counting, and calculation for reasons of agriculture and stock keeping.

In the ancient world, mathematics dealing with problems of geoscientific relevance flourished for the first time, for example, when Eratosthenes (276–195 BC) of Alexandria calculated the radius of the Earth. We also have evidence that the Arabs carried out an arc measurement northwest of Bagdad in the year 827 AD. Further key results of geomathematical research lead us from the Orient across the occidental Middle Ages to modern times. N. Copernicus (1473–1543) successfully made the transition from the Ptolemaic geocentric system to the heliocentric system. J. Kepler (1571–1630) determined the laws of planetary motion. Further milestones from a historical point of view are, for example, the theory of geomagnetism developed by W. Gilbert (1544–1608), the development of triangulation methods for the determination of meridians by T. Brahe (1547–1601) and W. Snellius (1580–1626), the laws of falling bodies by G. Galilei (1564–1642), and the basic theory on the propagation of seismic waves by C. Huygens (1629–1695). The laws of gravitation formulated by I. Newton (1643–1727) have taught us that gravitation decreases with

© The Author(s), under exclusive license to Springer Nature Switzerland AG 2019
W. Freeden et al., *An Invitation to Geomathematics*, Lecture Notes in Geosystems
Mathematics and Computing, https://doi.org/10.1007/978-3-030-13054-1_2

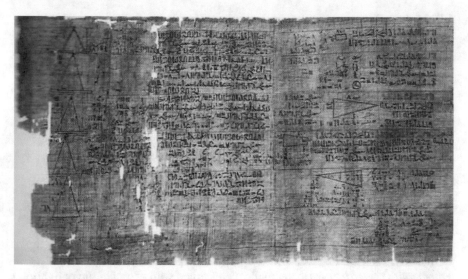

Fig. 2.1 Papyrus scroll containing indications of algebra, geometry, and trigonometry due to Ahmose (nineteenth century BC) [Department of Ancient Egypt and Sudan, British Museum EA 10057, London, Creative Commons Lizenz CC-BY-SA 2.0] taken from [113]

an increasing distance from the Earth. In the seventeenth and eighteenth century, France took over an essential role through the foundation of the Academy in Paris (1666). Successful discoveries include the theory of the isostatic balance of mass distribution in the Earth's crust by P. Bouguer (1698–1758), the calculation of the Earth's shape and especially of the pole flattening by P. L. Maupertuis (1698–1759) and A. C. Clairaut (1713–1765), and the development of the calculus of spherical harmonics by A. M. Legendre (1752–1833) and P. S. Laplace (1749–1829). The nineteenth century was marked by many contributions by C. F. Gauss (1777–1855). Especially important was the calculation of the lower Fourier coefficients of the Earth's magnetic field, the hypothesis of electric currents in the ionosphere, as well as the definition of the level set of the geoid (however, the term "geoid" was introduced by J. B. Listing (1808–1882), a disciple of C.F. Gauss). In 1849, G. G. Stokes (1819–1903) laid the foundation of the fundamental boundary value problems of physical geodesy (cf. [114]). At the end of the nineteenth century, the basic idea of the dynamo theory in geomagnetics was developed by B. Stewart (1851–1935), etc. This very incomplete list (which does not even include the last century) already shows that geomathematics may be understood as a cultural asset. Indeed, it is one of the large achievements of mankind from a historic point of view.

2.1 Geomathematics as Task and Objective

When mathematics yields good models of real phenomena, then mathematical reasoning can provide insight or predictions about nature. Accordingly, geomathematics deals with the qualitative and quantitative properties of the current or possible structures of the system Earth. It inspires concepts of scientific research concerning the system Earth, and it is simultaneously the force behind it.

The system Earth consists of a number of elements which represent individual systems themselves. The complexity of the entire system Earth is determined by interacting physical, biological, and chemical processes transforming and transporting energy, material, and information. It is characterized by natural, social, and economic processes influencing one another. In most instances, a simple theory of cause and effect is therefore completely inappropriate if we want to understand the system. We have to think in dynamical structures and to account for multiple, unforeseen, and of course sometimes even undesired effects in the case of interventions. Inherent networks must be recognized and used, and self-regulation must be accounted for. All these aspects require a type of mathematics which must be more than a mere collection of theories and numerical methods.

Mathematics dedicated to geosciences, i.e., *geomathematics*, deals with nothing more than the organization of the complexity of the system Earth.

Descriptive thinking is required in order to clarify abstract complex situations. We also need a correct simplification of complicated interactions, an appropriate system of mathematical concepts for their description, and exact thinking and formulations. Geomathematics has thus become the key science of the complex system Earth. Wherever there are data and observations to be processed, e.g., the diverse scalar, vectorial, and tensorial clusters of satellite data, we need mathematics. For example, statistics serves for noise reduction, constructive approximation for compression, and evaluation, the theory of special function systems yields georelevant graphical and numerical representations—there are mathematical algorithms everywhere. The specific task of geomathematics is to build a bridge between mathematical theory and geophysical as well as geotechnical applications.

The special attraction of this branch of mathematics is therefore based on the vivid communication between applied mathematicians more interested in model development, theoretical foundation, and the approximate as well as computational solution of problems, and geoengineers and geophysicists more familiar with measuring technology, methods of data analysis, implementation of routines, and software applications.

There is a very wide range of modern geosciences on which geomathematics is focused (see Fig. 2.2), not least because of the continuously increasing observation diversity. Simultaneously, the mathematical "toolbox" is becoming larger. A special feature is that geomathematics primarily deals with those regions of the Earth which are only insufficiently or not at all accessible for direct measurements (even by remote sensing methods). Mostly, a physical quantity is measured in the vicinity of

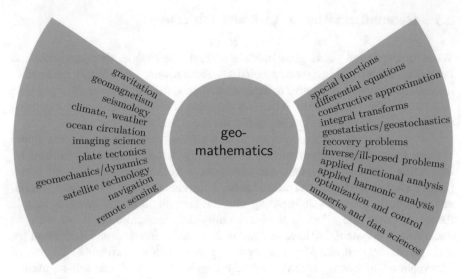

Fig. 2.2 Geomathematics, its range of fields, and its disciplines (in modified form from [47])

the Earth's surface, and it is then continued downward or upward by mathematical methods until one reaches the interesting depths or heights.

In conclusion, geomathematics has to deal essentially with what is meant by a *recovery problem* (RP). At present there exist a number of distinct approaches for the investigation and resolution of RPs. Indeed, during the last century a substantial amount of machinery from functional analysis, approximation theory, special function theory, potential theory, optimization, and numerical analysis, etc., has been brought to bear on the resolution and understanding of a recovery problem, and the interdisciplinary character of many recovery problems has emerged very clearly. So, the geomathematical schedule today is characterized by a twofold subdivision into typical parts, viz., sampling and inverse (ill-posed) problems on the one side and standard Euclidean as well as spherical framework on the other side.

Inverse problems (IPs) deal with determining for a given input–output system an input that produces an observed output, or of determining an input that produces a desired output (or comes as close to it as possible), often in the presence of noise. IPs arise in many branches of science and mathematics, including computer vision, natural language processing, machine learning, statistics, medical imaging (such as tomography and EEG/ERP), remote sensing, non-destructive testing, astronomy, Earth's system mathematics (especially involving, e.g., gravimetric, geomagnetic, and seismic exploration and satellite observational technology), and many other fields.

Sampling problems (SPs) have originally received considerable attention in the engineering community. The rudiments of sampling are covered in almost any engineering textbook on signal analysis. In the last three decades, however, with the surge of new techniques in analysis especially multiscale methods, sampling theory has started to take a prominent role within the traditional branches of mathematics, thereby leading to new discoveries in areas of geomathematics (see, e.g., [36, 38, 51]). Because of recent developments, for example, in satellite technology, there is a strong need to gather essential results of sampling in a unifying concept of geomathematically reflected recovery problems.

Signal Analysis/Processing (SA) deals with digital representations of signals and their analog reconstructions from digital representations. Techniques of functional analysis, computational and harmonic analysis play pivotal roles in this area.

Image Analysis and Processing are concerned with image refinement and recovery, and particularly include (geo)geophysics as well as medical imaging.

Moment problems deal with the recovery of a function or signal from its moments, and the construction of efficient stable algorithms for determining or approximating the function. Again this usually turns out to be an ill-posed problem.

Interrelated applications of inverse problems, signal analysis, and moment problems arise, in particular, in image analysis and recovery and in many areas of science and technology. Following Nashed [94], several decades ago the connections among all these areas (inverse problems, signal processing, and image analysis) were rather tenuous. Researchers in one of these areas were often unfamiliar with the techniques and relevance of the other two areas. The situation, however, has changed drastically in the last 20 years, and geomathematics certainly is a driving force behind this development. The common thread among inverse and sampling problems, signal analysis, and imaging is a canonical problem: *Recovering an object* (function, signal, and picture) from *partial or indirect information about the object.*

2.2 Geomathematics as Challenge

From a scientific and technological point of view, the past twentieth century was a period with two entirely different faces concerning research and its consequences. The first two-thirds of the century were characterized by a movement towards a seemingly inexhaustible future of science and technology; they were marked by the absolute belief in technical progress which would make everything achievable in the end. Up to the 1960s, mankind believed to have become the master of the Earth (note that, in geosciences as well as other sciences, to master is also a synonym for to understand). Geoscience was able to understand plate tectonics on the basis of Wegener's theory of continental drift, geoscientific research began to deal with the Arctic and Antarctic, and man started to conquer the universe by satellites, so that for the first time in mankind's history the Earth became measurable on a global scale, etc. Then, during the last third of the past century, there was a growing skepticism as to the question whether scientific and technical progress had really brought us forth and whether the effects of our achievements were responsible. As a consequence of the specter of a shortage in raw materials (mineral oil and natural gas reserves), predicted by the Club of Rome, geological/geophysical research with the objective of exploring new reservoirs was stimulated during the 1970s. Moreover, the last two decades of the century have sensitized us for the global problems resulting from our

Fig. 2.3 Geoid (more concretely, GFZ-EIGEN-CG01C geoid (2005))

behavior with respect to climate and environment. Our senses have been sharpened as to the dangers caused by the forces of nature, from earthquakes and volcanic eruptions, to the temperature development and the hole in the ozone layer, etc. Man has become aware of his environment.

Satellites orbiting the Earth have confirmed that the planet is not a simple sphere we often imagine it to be. It is more like of a potato (see Fig. 2.3). In fact, based on gravitational measurements, satellite geodesy has yielded a colourful map portraying the geoid in potato form, characterizing the shape of the Earth's surface, if it were entirely covered by water and influenced by gravity alone.

Similarly, the image of the Earth used by oceanographers as a huge potato drenched by rainfall is also not a false one. The humid layer on this potato, maybe only a fraction of a millimeter thick, is the ocean. The entire atmosphere hosting the weather and climate events is only a little bit thicker. Flat bumps protruding from the humid layer represent the continents. The entire human life takes place in a very narrow region of the outer peel (only a few kilometers in vertical extension). However, the basically excellent comparison of the Earth with a huge potato does not give explicit information about essential ingredients and processes of the system Earth, for example, magnetic field, deformation, wind and heat distribution, ocean currents, internal structures, etc.

In our twenty-first century, geoproblems currently seem to overwhelm the scientific programs and solution proposals. *"How much more will our planet Earth be able to take?"* has become an appropriate and very urgent question. Indeed, there has been a large number of far-reaching changes during the last few decades, e.g., species extinction, climate change, formation of deserts, ocean currents, structure of the atmosphere, transition of the dipole structure of the magnetic field to a quadrupole structure, etc. These changes have been accelerated dramatically. The main reasons for most of these phenomena are the unrestricted growth in the industrial societies (population and consumption, especially of resources, territory, and energy) and severe poverty in the developing and newly industrialized countries. The dangerous aspect is that extreme changes have taken place within a very short time; there has been no comparable development in the dynamics of the system

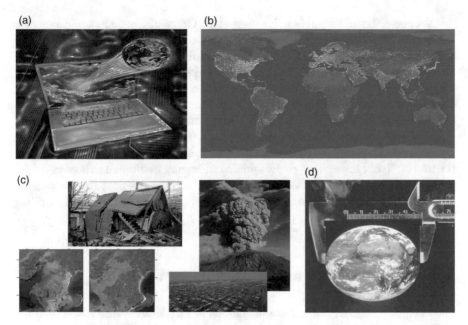

Fig. 2.4 Four significant reasons for the increasing importance of geomathematics. (**a**) Modern high speed computers are entering more and more all geodisciplines. (**b**) There exists a growing public concern about the future of our planet, its climate, its environment, and about an expected shortage of natural resources. (**c**) There is a strong need for strategies of protection against threats of a changing Earth. (**d**) There is an exceptional situation of getting data of better and better quality

Earth in the past. Changes brought about by human beings are much faster than changes due to natural fluctuations. Besides, the last financial crisis showed that a model of affluence (which holds for approximately 1 billion people) cannot be transferred globally to about 8 billion people. Massive effects on mankind such as migration, etc., are inevitable. The appalling situation is that the geoscientific problems collected over the decades must now all be solved simultaneously. Interdisciplinary solutions including intensive mathematical research are urgently required as answers to an increasingly complex world. Geomathematics is absolutely essential for a sustainable development in the future (Fig. 2.4).

However, the scientific challenge does not only consist of increasing the leading role of mathematics within the current "scientific consortium Earth." The significance of the subject "Earth" must also be acknowledged (again) within mathematics itself, so that mathematicians will become more enthusiastic about it. Up to now, it has become usual and good practice in application-oriented mathematical departments and research institutions to present applications in technology, economy, finances, and even medicine as being very career-enhancing for young scientists. Geomathematics can be integrated smoothly into this phalanx with current subjects like climate change, exploration, navigation, and so on.

Of course, basic research is indispensable. Geomathematics should not hide behind the other geosciences! Neither should we wait for the next natural hazard to take place! Now is the time to turn towards georelevant applications. The Earth as a complex, however, limited system (with its global problems concerning climate, environment, resources, and population) needs new political strategies. Consequently, these will step by step also demand changes in research due to our modified concept of "well-being" (e.g., concerning milieu, health, work, independence, financial situation, security, etc.). The time has come to realize that geomathematics is also indispensable as a constituting discipline within universities (instead of "ivory tower like" parity thinking following traditional structures).

All in all, mathematics should be one of the leading sciences for the solution of these complex and economically very interesting problems, instead of fulfilling mere service functions.

2.3 Geomathematics as Interdisciplinary Discipline

The canonical outcome of our considerations is that *mathematics as an interdisciplinary science* can be found in almost every area of our lives. Mathematics is closely interacting with almost every other science, even medicine and parts of the arts (*"mathematization of sciences"*). The use of computers allows for the handling of complicated models for real data sets. Modeling, computation, and visualization yield reliable simulations of processes and products. Mathematics is the "raw material" for the models and the essence of each computer simulation. As the key technology, it translates the images of the real world to models of the virtual world, and vice versa (cf. Fig. 2.5).

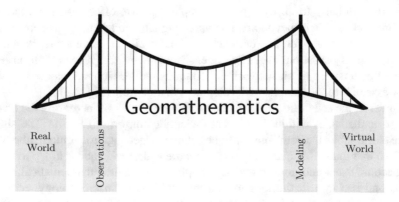

Fig. 2.5 Geomathematics as a key technology bridging the real and virtual world (cf. [47])

As a matter of fact, the special importance of mathematics as an interdisciplinary science has been acknowledged increasingly within the last few years in technology, economy, and commerce. However, this process does not remain without effects on mathematics itself. New mathematical disciplines, such as scientific computing, financial and business mathematics, industrial mathematics, biomathematics, and also geomathematics, have complemented the traditional disciplines. Interdisciplinarity also implies the interdisciplinary character of mathematics at school. Relations and references to other disciplines (especially informatics, physics, chemistry, biology, and also economy and geography) become more important, more interesting, and more expandable. Problem areas of mathematics become explicit and observable, and they can be visualized. Of course, this undoubtedly also holds for the system Earth.

As already pointed out, geomathematics is closely interconnected with geophysics and geoinformatics:

Geophysics is concerned with the physical processes and physical properties of the Earth and its surrounding space environment, and the use of physical methods for their analysis. In the past, the term geophysics usually refers only to the geologically oriented applications: Earth's shape; its gravitational and magnetic fields; its internal structure and composition; and its dynamics and their surface expression in plate tectonics, the generation of magmas, volcanism, and rock formation. However, modern geophysics organizations use a broader definition that includes the water cycle including snow and ice; fluid dynamics of the oceans and the atmosphere; electricity and magnetism in the ionosphere and magnetosphere; and solar–terrestrial relations (for more details see https://en.wikipedia.org/wiki/Geophysics). The historical development of geophysics has been motivated by two factors: One of these is the research of humankind related to Planet Earth and its several components, its events, and its problems. The second is the economical usage of Earth's resources (ore deposits, petroleum, water resources, etc.) and Earth-related hazards such as earthquakes, volcanoes, tsunamis, tides, and floods. As a consequence, the research areas in geomathematics and geophysics are coincident and concurrent. However, the affiliation is different. Geophysics is a discipline of natural science. The first known use of the word geophysics was by J. Fröbel in 1834 (in German). It was used occasionally in the next few decades, but did not catch on until journals devoted to the subject began to appear, beginning with "Beiträge zur Geophysik" in 1887. The future "Journal of Geophysical Research" was founded in 1896 with the title "Terrestrial Magnetism." In 1898, a Geophysical Institute was founded at the University of Göttingen, and Emil Wiechert (1861–1928) became the world's first Chair of Geophysics. An international framework for geophysics was provided by the founding of the International Union of Geodesy and Geophysics in 1919. The twentieth century was a revolutionary age for geophysics. As an international scientific effort between 1957 and 1958, the "International Geophysical Year" was one of the most important for scientific activity of all disciplines of geophysics: cosmic rays, geomagnetism, gravity, ionospheric physics, longitude and latitude determinations (precision mapping), meteorology, oceanography, seismology, and solar activity (cf. https://en.wikipedia.org/wiki/History_of_geophysics).

Geoinformatics basically differs from geomathematics. While geomathematics also deals with the further development of the language itself, geoinformatics, however, concentrates on the design and architecture of processors and computers, databases and programming languages, etc., in a georeflecting environment. In geomathematics, computers do not represent the objects to be studied, but instead represent technical auxiliaries for the solution of mathematical problems of geo-reality (for more details see, e.g., https://fatwaramdani.wordpress.com/2016/02/19/what-is-geoinformatics/).

All in all, the terms "geophysics" and "geoinformatics" were coined many decades ago and are now recognized and well-defined disciplines.

Statistics (sometimes seen as a basic subdiscipline of mathematics) is generally devoted to the analysis and interpretation of uncertainties caused by limited sampling of a property under study. In consequence, the focus of *geostatistics* is the development and statistical validation of models to describe the distribution in time and space of Earth sciences phenomena. *Geostatics* was originated in the mining industry. In the early 1950s, when standard statistics were found unsuitable for estimating disseminated ore reserves, D.G. Krige (1919–2013), a South African mining engineer, and H.S. Sichel (1915–1995), a statistician, developed a new estimation method. Later, the petroleum industry became interested in geostatistics to improve the modeling of the geometry of oil reservoirs using data from a small number of wells combined with more detailed data from seismic surveys. It's now used for all types of environmental study—air, water, and ground—for which there is an increasing demand for justifiable maps where the uncertainty is quantified (for more details see, e.g., https://en.wikipedia.org/wiki/Geostatistics).

In 1974, Agterberg [2] already used the term "geomathematics" for the title of his book, but he did that in the line of geostatistics for purposes of the mining industry. In our understanding, modern geomathematical publications act on a broader spectrum. Today (more specifically, since the year 2010), geomathematics runs its own journal "GEM International Journal on Geomathematics" as the broad forum for all mathematical "planet Earth obligations" (for more details see the Appendix A at the end of this work).

2.4   Geomathematics as Generically Constituted Discipline

No branch of mathematics has influenced the general scientific thinking more than the mathematical theories originated from fields of physics such as mechanics. In the mathematical community, there is no doubt that the philosophy and the outcome of this influence to "reality" is the most characteristic feature of our technical civilization. However, a statement of such generality, though accepted by most mathematicians, is most likely questioned by many physicists. The reason for the difference of opinion is the lack of a generally accepted definition of what is indicated by the mathematical way of thinking.

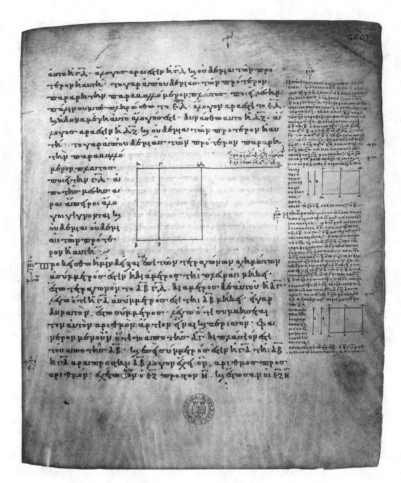

Fig. 2.6 Euklid, Elemente 10, Appendix in der 888 geschriebenen Handschrift Oxford, Bodleian Library, MS. D'Orville 301, fol. 268r

In what follows, influenced by Müller's note [89], we first make the attempt to show the different tendencies of the line of thoughts in the interrelation and interaction of mathematical and physical sciences. In a second step it also characterizes the specific interplay of geophysics and geomathematics.

Seen from the historic point of view (cf. [47]), one of the most important contributions to mathematics is *geometry* (from the Ancient Greek γεωμετρία geo-"Earth," -metron "measurement"). The earliest recorded beginnings of geometry (Fig. 2.6) can be traced to ancient Mesopotamia and Egypt in the 2nd millennium BC. Early geometry was a collection of empirically discovered principles concerning distances, lengths, angles, areas, surfaces, and volumes, which were developed to meet some practical need in surveying and various crafts. The earliest known texts (cf. Fig. 2.1) on geometry are the Egyptian Rhind Papyrus (2000–1800 BC), Moscow

Fig. 2.7 Euclid's elements is by far the most famous mathematical work of classical antiquity, and also has the distinction of being the world's oldest continuously used mathematical textbook. Little is known about the author, beyond the fact that he lived in Alexandria around 300 BC. The main subjects of the work are geometry, proportion, and number theory

Papyrus (roughly 1890 BC), and the Babylonian clay tablets (roughly 1900 BC). Geometry arose independently in a number of early cultures as a practical way for dealing with lengths, areas, and volumes. Geometry began to see elements of formal mathematical science emerging in the West as early as the sixth century BC.

By the third century BC, geometry was put into an axiomatic form by Euclid (cf. Fig. 2.7), whose treatment, *Euclid's Elements*, set a standard for many centuries to follow. It brought the heritage of mathematics from the antiquity to our time. Some centuries later, the Greeks themselves replaced the term "geometry" (cf. [45]), which had meanwhile lost the original meaning of "Earth's measuring" by "geodesy" as a new meaning of an abstract theory of the "Earth's shape," while geometry now reflected the mathematical rigor through its axiomatic method. In fact, it is the earliest example of the format still used in mathematics today that of definition, axiom, theorem, and proof. Although most of the contents of the "Elements" were already known, Euclid arranged them into a single, coherent logical framework. The "Elements" were known to all educated people in the West until the middle of the twentieth century and its contents are still taught in geometry classes today. The demand for intellectual rigor, the wealth of knowledge, and the high standard of thinking prove the great significance of mathematics in classical Greece.

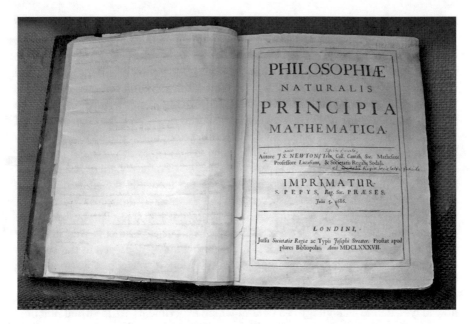

Fig. 2.8 Sir Isaac Newton's own first edition copy of his Philosophiae Naturalis Principia Mathematica with his handwritten corrections for the 20th edition. The first edition was published under the imprint of Samuel Pepys who was president of the Royal Society. By the time of the second edition, Newton himself had become president of the Royal Society, as noted in his corrections. It has been digitized by Cambridge University Library and can be seen in the Cambridge Digital Library along with other original works by Isaac Newton. The book can be seen in the Wren Library of Trinity College, Cambridge

In this respect, it should be mentioned that Islamic scientists preserved Greek ideas and expanded them during the Middle Ages. By the early seventeenth century, geometry had been put on a solid analytic footing by mathematicians such as René Descartes (1596–1650) and Pierre de Fermat (1607–1665). Since then, and into modern times, geometry has expanded into non-Euclidean geometry, describing spaces that lie beyond the normal range of human experience. Hence, while geometry has evolved significantly throughout the years, there are some general concepts that are more or less fundamental. These include the concepts of points, lines, angles, curves, planes, and surfaces, as well as the more advanced notions of manifolds and topology or metric.

Historians very often use the term modern (also, in geosciences) for the period starting with the Italian Renaissance. Scientists probably call Isaac Newton (1643–1727) the first modern scientist. The emphasis on a systematic study of the laws such as the Newtonian approach in *"Philosophiae Naturalis Principia Mathematica"* (cf. Fig. 2.8) may be regarded as the most obvious characteristic of modern times. It also initiated a new branch of mathematics, called *analysis*, which from its very beginning is so closely related to physics, for example, in the field of differential equations, that some people find it difficult to say, where mathematics starts and physics ends. The history of the interchange of ideas between the areas of mathematics and physics is a fascinating period in the history of the scientific thought. Unfortunately, it is so intricate that its complete account is still to be investigated. Nevertheless, there is a large variety of basically different ideas and concepts and an increasing change of the aspects involved in the problems. So it may be concluded that one great contribution of mathematics to modern scientific progress is the setting of the pattern for the theories describing the laws of the real world ("reality"). Without mathematical theory providing "virtuality" there is no science in a modern sense.

Apart from the mathematical aspect of the value of a theory, there is a practical side to which it is even more important. Once the theory is recognized to be in accordance with an experience, it renders the experiment unnecessary. This, of course, is very important where experiments are not possible. The best example of a geodetic field for which we could not carry out experiments was probably astronomical geodesy. So it was not surprising that a new approach to the features of the "real world" had its striking success when it was possible to deduce Kepler's empirical law on the movements of planets from Newton's general law of mechanics. In fact, the same rules also apply to today's space research, and nowadays satellite technology is a field in which experiments are the usual practice.

Looking at the developments of the interrelations between mathematics, engineering, and physics during the last century we are led to the following conclusion (cf. [89]): The aspects of mathematically reflected problems of physics as well as engineering brought about that in these fields more proofs are constructive than is the case in many other fields of mathematics. However, the aspect of a mathematical theory, the way of looking at its problems, is not only changed by contacting outside areas. A theory may get innovative incentives from a change of emphasis and the discovery of a new framework based on mathematical equipment exclusively. For example, the theory of vector spaces and linear operators which originated from the Fredholm theory of boundary value problems is now regarded as the common heading for a large class of problems in analysis. In fact, they created a new mathematical discipline, today called *functional analysis*. In this respect, Müller [89] states:

> It achieves a general formulation of a variety of problems, brings out the mathematical structure and thus simplifies the theory by emphasizing the essential. By reducing the assumptions to a minimum, it increases the range of applications. As a matter of fact, there can be no doubt that the historic core of the discipline "functional analysis" is the Fredholm theory of integral equations and the Hilbert theory of eigenfunctions. These approaches provided the basis, that was then simplified and extended by more abstract concepts.

Nowadays, operator theory of functional analysis in the manifestation of regularization methods even helps to handle *ill-posed inverse problems* for a large variety of applications. In geodesy, for example, two essential areas for ill-posed inverse problem types may be specified, namely, "downward continuation" of satellite data to the Earth's surface and "downward computation" of terrestrial data to the Earth's interior. The rationale in most methods for resolution (approximate solvability) of ill-posed inverse problems is to construct a "solution" that is acceptable physically as a meaningful approximation and is sufficiently stable from the computational standpoint; hence, an emphasis is put on the distinction between "solution" and "resolution" (see, e.g., [91–99]). The main dilemma of modeling of ill-posed problems (IPP) is that the closer the mathematical model describes the IPP, the worse is the "condition number" of the associated computational problem (i.e., the more sensitive to errors). For ill-posed problems, the difficulty is to bring additional information about the desired solution, compromises, or new outlooks as aids to the resolution of IPP. It is conventional to use the phrase "regularization of an ill-posed problem" to refer to various approaches to circumvent the lack of continuous dependence (as well as to bring about existence and uniqueness if necessary). Roughly speaking, this entails an analysis of an IPP via an analysis of an associated well-posed problem, i.e., a system (usually a sequence or a family) of well-posed problems, yielding meaningful answers to the IPP. We distinguish three aspects of regularization:

- strategy of resolution and reconstruction,
- regularization-approximation schemata, and
- regularization algorithms.

One of the purposes of geomathematical work is to dramatize this delineation with reference to specific methods and results.

The *strategy of resolution and reconstruction of ill-posed problems* involves one or more of the following intuitive ideas (see, e.g., [37, 41, 92, 93]):

- change the notion of what is meant by a solution (e.g., ε-approximate solution: $\|Au - y\| \leq \varepsilon$, where $\varepsilon > 0$ is prescribed; quasi-solution: $\|Au - y\| \leq \|Ax - y\|$ for all $x \in \mathcal{M}$, a prescribed subset of the domain of A; least-squares solution; pseudoinverse, etc.),
- modify the operator equation or the problem itself,
- change the spaces and/or topologies, and
- specify the type of involved noise ("strong" or "weak" noise, see [18] and the references therein).

The philosophy of resolution leads to the use of algebraic methods versus function space methods, statistical versus deterministic approaches, strong versus weak noise, etc. A *regularization-approximation scheme* refers to a variety of methods such as Tikhonov's regularization, projection methods, multiscale methods, iterative approximation, etc., that can be applied to ill-posed problems. These schemes turn into algorithms once a resolution strategy can be effectively implemented. Unfortunately, this requires a determination of a suitable value of a certain parameter associated with the scheme (e.g., regularization parameter, mesh size, dimension of subspace in the projection scheme, specification of the level of a scale space, classification of noise, etc.). This is not a trivial problem (see [45] for a more detailed explanation seen from physical geodesy), since it involves a trade-off between accuracy and numerical stability, a situation that does not usually arise in well-posed problems.

From the standpoint of mathematical and numerical analysis one can roughly group "regularization methods" into several categories (see, e.g., [4–6, 18, 19, 37, 74, 91–93, 122] and the references therein):

1. Regularization methods in function spaces is one category. This includes Tikhonov-type regularization, the method of quasi-reversibility, the use for certain function spaces such as scale spaces in multi-resolutions, the method of generalized inverses (pseudoinverses) in reproducing kernel Hilbert spaces, and multiscale wavelet regularization.
2. Resolution of ill-posed problems by "control of dimensionality" is another category. This includes projection methods and discretization moment–discretization schemes. The success of these methods hinges on the possibility of obtaining approximate solution while keeping the dimensionality of the finite-dimensional problem within the "range of numerical stability." It also hinges on deriving error estimates for the approximate solutions that is crucial to the control of the dimensionality.
3. A third category is iterative methods which can be applied either to the problem in function spaces or to a discrete version of it. The crucial ingredient in iterative methods is to stop the iteration before instability creeps into the process. Thus iterative methods have to be modified or accelerated so as to provide a desirable accuracy by the time a stopping rule is applied.
4. A fourth category is filter methods. Filter methods refer to procedures where, for example, values producing highly oscillatory solutions are eliminated. Various "low-pass" filters can, of course, be used. They are also crucial for the determination of a stopping rule. Mollifiers are known in filtering as smooth functions with special properties to create sequences of smooth functions approximating a non-smooth function or a singular function.
5. The original idea of a mollifier method (see, e.g., [19] and the references therein) is interested in the solution of an operator equation, but we realize that the problem is "too ill-posed" for being able to determine the (pseudo)inverse accurately. Mollifiers are known as smooth functions with special properties to create sequences of smooth functions approximating a non-smooth function. Thus, we compromise by changing the problem into a more well-posed one, namely, that of trying to determine a mollified version of the solution. The heuristic motivation is that the trouble usually comes from high-frequency components of the data and of the solution, which are damped out by a mollification.
6. The root of the Backus–Gilbert method (BG-method) was geophysically motivated (cf. [4–6]). The characterization involved in the model is known as a moment problem in the mathematical literature. The BG-method can be thought of as resulting from discretizing an integral equation of the first kind. Where other regularization methods, such as the frequently used Tikhonov regularization method (see, e.g., [40] and the references therein), seek to impose smoothness constraints on the solution, the BG-method instead realizes stability constraints. As a consequence, the solution is varying as little as possible if the input data were resampled multiple times. The common feature between mollification and

the BG-method is that an approximate inverse is determined independently from the right-hand side of the equation.

Generally, it should be noted that there are two main *trends in mathematics*, which cannot exist one without the other: The first is to discover the laws for relations between mathematical concepts and the second is to make these structures by a process of generalization and extension as applicable and powerful as possible. As a consequence, geomathematics as a generic discipline is interpenetrated by an alternative that may be characterized as follows:

- On the one hand, the range of mathematics contains many results which are noted for their generality like functional analysis and its manifestation in ill-posed problems.
- On the other hand, many of the famous unsolved problems are of a very special nature. All these problems deal with questions of great individuality in the sense that the theories generally developed so far do not yet contain the information which is necessary to elucidate the specific ingredients of the conjectures in question.

2.5   Geomathematics as Spherical Solution Method

Up to now, trial functions for the description of geoscientifically relevant parameters have been frequently based on the (almost) spherical geometry of the Earth. Although a mathematical formulation in a spherical context may be a restricted simplification, it is at least acceptable for a large number of problems. In fact, ellipsoidal nomenclature is much closer to geophysical and/or geodetic purposes at least when obligations of the Earth's centrifugal potential have to be considered (cf. [55]), but the computational and numerical amount of work usually is a tremendous obstacle for gravitational purposes.

Next we almost literally follow the approach presented in the introductory part of the monograph [51] to explain the solution potential in geomathematics by specifically spherical equipment (for more details and references see also [51]): There is a *palette of signals* to be studied on the sphere. For instance, the space variation of a spherical signal is fundamental for many applications. However, if we are interested in gaining a deeper understanding of the space variation, it is often advantageous to study the signal in different representations. For example, the signal can be obtained from a complete system of polynomials, e.g., spherical harmonics, providing a spectral (frequency) representation. From a mathematical point of view there is an infinite number of ways this can be done.

Even more, the idea that a discontinuous signal (function) on the sphere may be expressed as a sum of arbitrarily often differentiable polynomials on the sphere turned out to be one of the great innovations since the time of Laplace, Legendre, and Gauss.

Energy of a Signal   How much energy a spherical signal has and how much energy it takes to produce are central problems in geosciences. Signal analysis has been extended to many diverse types of data with different understanding of energy. Today, the usual understanding of the total energy of a signal F is achieved by the "continuous summation," i.e., integration over all space (the unit sphere) $\mathbb{S}^2$ of the pointwise "fractional (pointwise) energy" $|F(\xi)|^2$, $\xi \in \mathbb{S}^2$, in the form

$$\|F\|_{L^2(\mathbb{S}^2)}^2 = \int_{\mathbb{S}^2} |F(\xi)|^2 \, dS(\xi), \tag{2.1}$$

where dS is the surface element.

Spherical Harmonics The space $L^2(\mathbb{S}^2)$ of all signals (functions) having finite energy may be appropriately characterized by certain systems of restrictions of homogeneous harmonics polynomials to the sphere, in fact, leading canonically to a spherical harmonics system constituting a Hilbert space. The polynomial structure of spherical harmonics $\{Y_{n,k}\}_{n=0,1,\ldots,k=-n,\ldots,n}$ in the Hilbert space $L^2(\mathbb{S}^2)$ has a tremendous advantage. First, spherical harmonics of different degrees are orthogonal (in the topology implied by (2.1)). Second, the space

$$\mathrm{Harm}_n = \mathrm{span}_{k=-n,\ldots,n} Y_{n,k} \tag{2.2}$$

of spherical harmonics of degree (frequency) n is finite-dimensional. Its dimension is given by $\dim(\mathrm{Harm}_n) = 2n + 1$, so that

$$\mathrm{Harm}_{0,\ldots,m} = \mathrm{span}_{\substack{n=0,\ldots,m \\ k=-n,\ldots,n}} Y_{n,k} = \bigoplus_{n=0}^{m} \mathrm{Harm}_n \tag{2.3}$$

implies $\dim(\mathrm{Harm}_{0,\ldots,m}) = \sum_{n=0}^{m} 2n + 1 = (m+1)^2$. The basis property of $\{Y_{n,k}\}_{n=0,1,\ldots,k=-n,\ldots,n}$ in the space $L^2(\mathbb{S}^2)$ of finite-energy signals is equivalently characterized by the completion of the orthogonal direct sum $\bigoplus_{n=0}^{\infty} \mathrm{Harm}_n$, i.e.,

$$L^2(\mathbb{S}^2) = \overline{\bigoplus_{n=0}^{\infty} \mathrm{Harm}_n}^{\|\cdot\|_{L^2(\mathbb{S}^2)}}. \tag{2.4}$$

This is the natural reason why spherical harmonics expansions are essential tools not only in the scalar theory of gravitational and geomagnetic potentials, but also in vectorial research areas of fields, e.g. (geostrophic) ocean circulation and geodeformation (Fig. 2.9).

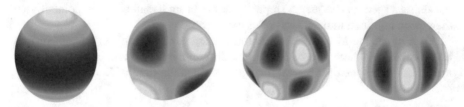

Fig. 2.9 Spherical harmonics of low degrees

Spectral analysis in terms of spherical harmonics $\{Y_{n,k}\}_{n=0,1,\ldots,k=-n,\ldots,n}$ has led to the discovery of basic laws of nature. It allows us to understand the composition and ingredients of features of the Earth (for more details about space and frequency description of (one-dimensional) signals see, e.g., [14] and the references therein).

The formalism of a spherical harmonics sampling system is essentially based on the following principles (cf. [46]):

1. The spherical harmonics are obtainable in a twofold way, namely, as restrictions of three-dimensional homogeneous harmonic polynomials or intrinsically on the unit sphere $\mathbb{S}^2$ as eigenfunctions of, e.g., the Beltrami operator or certain pseudodifferential operators.
2. The Legendre kernels (polynomials) are obtainable as the outcome of sums extended over a maximal horizontal orthonormal system of spherical harmonics (cf. Fig. 2.10) of degree (i.e., frequency) n.
3. The Legendre kernels are rotation-invariant with respect to orthogonal transformations (leaving one point of the unit sphere $\mathbb{S}^3$ fixed).

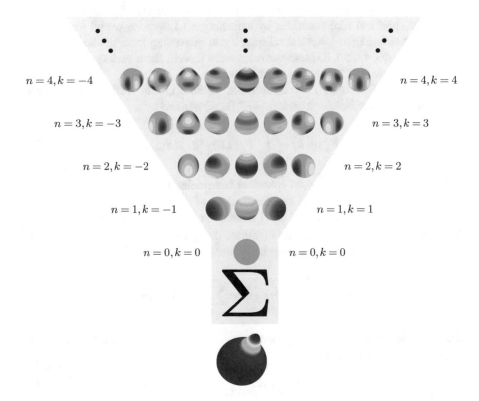

Fig. 2.10 Spacelimited zonal kernel generated as weighted infinite sum over spherical harmonics (cf. [46, 51])

4. Each Legendre kernel implies an associated Funk–Hecke formula that determines the constituting features of the convolution (filtering) of a square-integrable field against the Legendre kernel.
5. The orthogonal Fourier expansion of a square-integrable field is the sum of the convolutions of the field against the Legendre kernels being extended over all frequencies.

In fact, the theory of spherical harmonics provides a powerful spectral framework to unify, review, and supplement the different approaches in spaces over the unit sphere $\mathbb{S}^2$ in Euclidean space $\mathbb{R}^3$, where distance (norm) and angle are at hand in suitable reference (pre-)Hilbert spaces. The essential tools in these (pre-)Hilbert spaces are the Legendre functions, used in orthogonal Fourier expansions and endowed with rotational invariance. The coordinate-free construction yields a number of formulas and theorems that previously were derived only in problem-affected coordinate (more precisely, polar coordinate) representations. As a consequence, any kind of singularity is avoided at points being fixed under orthogonal transformations. Finally, the transition from the scalar to the vectorial as well as the tensorial case opens new promising perspectives of constructing important zonal classes of spherical trial functions by summing up Legendre kernel expressions, thereby providing (geo-)physical relevance and increasing local applicability (see, e.g., [30–32, 46, 49, 50, 83] and the references therein for a variety of aspects on constructive spherical harmonics approximation).

Any signal $F \in L^2(\mathbb{S}^2)$ can be split (cf. Table 2.1) into "orthogonal contributions" involving the Fourier transforms $F^\wedge(n, k)$ defined by

$$F^\wedge(n, k) = \int_{\mathbb{S}^2} F(\xi) Y_{n,k}(\xi) \, dS(\xi), \tag{2.5}$$

in terms of $L^2(\mathbb{S}^2)$-orthonormal spherical harmonics $\{Y_{n,k}\}_{n=0,1,\ldots,k=-n,\ldots,n}$. The total energy of a signal should be independent of the method used to calculate it. Hence, $\|F\|^2_{L^2(\mathbb{S}^2)}$ as defined by (2.1) should be the sum of $(F^\wedge(n, k))^2$ over all frequencies. So, Parseval's identity identifies the spatial energy of a signal with the spectral energy, decomposed orthogonally into single frequency contributions

$$\|F\|^2_{L^2(\mathbb{S}^2)} = \langle F, F \rangle_{L^2(\mathbb{S}^2)} = \sum_{n=0}^{\infty} \sum_{k=-n}^{n} \left(F^\wedge(n, k) \right)^2.$$

This explains why the (global) geosciences work more often with the frequency energy, i.e., *amplitude spectrum*

$$\left\{ F^\wedge(n, k) \right\}_{n=0,1,\ldots,k=-n,\ldots,n} \tag{2.6}$$

Table 2.1 Fourier expansion of $L^2(\mathbb{S}^2)$-functions (cf. [51])

Spherical harmonics $\{Y_{n,k}\}_{n=0,1,\ldots,k=-n,\ldots,n}$ as polynomial system on the unit sphere
$$\mathbb{S}^2 \subset \mathbb{R}^3$$

Orthonormality and invariance	addition theorem

$\downarrow$

One-dimensional Legendre polynomial P_n:
$$P_n(\xi \cdot \eta) = \tfrac{4\pi}{2n+1} \sum_{k=1}^{2n+1} Y_{n,k}(\xi) Y_{n,k}(\eta), \quad \xi, \eta \in \mathbb{S}^2$$

Convolution by the Legendre kernel	Funk–Hecke formula

$\downarrow$

Legendre transform of F:
$$(P_n * F)(\xi) = \tfrac{2n+1}{4\pi} \int_{\mathbb{S}^2} P_n(\xi \cdot \eta) F(\eta)\, dS(\eta) = \sum_{k=-n}^{n} F^\wedge(n, k) Y_{n,k}(\xi), \quad \xi \in \mathbb{S}^2$$

Fourier	coefficients

$\downarrow$

Fourier coefficients of $F \in L^2(\mathbb{S}^2)$:
$$F^\wedge(n, k) = \int_{\mathbb{S}^2} F(\xi) Y_{n,k}(\xi) dS(\xi)$$

Superposition over frequencies	orthogonal series expansion

$\downarrow$

Fourier series of $F \in L^2(\mathbb{S}^2)$:
$$F(\xi) = \sum_{n=0}^{\infty} \tfrac{2n+1}{4\pi} \int_{\mathbb{S}^2} P_n(\xi \cdot \eta) F(\eta)\, dS(\eta) = \sum_{n=0}^{\infty} \sum_{k=-n}^{n} F^\wedge(n, k) Y_{n,k}(\xi), \quad \xi \in \mathbb{S}^2$$

than with the original space signal $F \in L^2(\mathbb{S}^2)$. As a consequence, the "inverse Fourier transform"

$$F = \sum_{n=0}^{\infty} \sum_{k=-n}^{n} F^\wedge(n, k) Y_{n,k} \tag{2.7}$$

allows the geoscientists to think of the function (signal) F as weighted superpositions of "wave functions" $Y_{n,k}$ corresponding to different frequencies.

In this respect, one can think of measurements as operating on an "input signal" F to produce an output signal $G = \Lambda F$ (cf. [33, 34]), where Λ is an operator acting on $L^2(\mathbb{S}^2)$. Fortunately, large portions of interest can be well approximated by pseudodifferential operators. If Λ is such an operator on $L^2(\mathbb{S}^2)$, this means that

$$\Lambda Y_{n,k} = \Lambda^\wedge(n, k) Y_{n,k}, \quad n = 0, 1, \ldots, k = -n, \ldots, n, \tag{2.8}$$

where, in geodesy and geophysics, the "symbol" $\{\Lambda^\wedge(n, k)\}_{n \in \mathbb{N}_0, k=-n,\ldots,n,}$ usually is a sequence of real values independent of the order k, i.e., $\Lambda^\wedge(n, k) = \Lambda^\wedge(n)$ for all n . A pseudodifferential operator Λ satisfying

$$\Lambda^{\wedge}(n, k) = \Lambda^{\wedge}(n) \tag{2.9}$$

for all n is called *rotation-invariant* (or *isotropic*). Equation (2.8) allows the interpretation that the spherical harmonics are the eigenfunctions of the operator Λ. Different pseudodifferential operators Λ are characterized by their eigenvalues $\Lambda^{\wedge}(n)$. Moreover, the amplitude spectrum $\{G^{\wedge}(n, k)\}$ of the response of Λ is described in terms of the amplitude spectrum of functions (signals) by a simple multiplication by the "transfer" function $\Lambda^{\wedge}(n)$.

Signal Band and Space Limitation Physical devices do not transmit spherical harmonics of arbitrarily high frequency without severe attenuation. The transfer function $\Lambda^{\wedge}(n)$ usually tends to zero with increasing n. It follows that the amplitude spectra of the responses (observations) to functions (signals) of finite energy are also negligibly small beyond some finite frequency. Thus, both because of the frequency limiting nature of the devices used, and because of the nature of the "transmitted signals," the geoscientist is soon led to consider bandlimited functions. These are the functions $F \in L^2(\mathbb{S}^2)$ whose "amplitude spectra" vanish for all $n \geq N$ (for some fixed $N \in \mathbb{N}_0$).

A bandlimited function $F \in L^2(\mathbb{S}^2)$ can be written as a finite Fourier series. So, any function F of the form

$$F = \sum_{n=0}^{N} \sum_{k=-n}^{n} F^{\wedge}(n, k) Y_{n,k} \tag{2.10}$$

is said to be *bandlimited with the band N*, if $F^{\wedge}(N, k) \neq 0$ for at least one k.

If there exists a region $\Gamma \subsetneq \mathbb{S}^2$ such that $F \in L^2(\mathbb{S}^2)$ vanishes on $\mathbb{S}^2 \backslash \Gamma$, F is said to be spacelimited (locally supported). $F \in L^2(\mathbb{S}^2)$ is called *spacelimited (locally supported) with spacewidth $\varrho \in (-1, 1)$* around an axis $\eta \in \mathbb{S}^2$, if the function F vanishes for some $\varrho \in (-1, 1)$ on the set of all unit vectors $\xi \in \mathbb{S}^2$ with $-1 \leq \xi \cdot \eta \leq \varrho$ (where ϱ is the largest number for which this is the case).

Bandlimited functions are infinitely often differentiable everywhere. Moreover, it is clear that any bandlimited function F is an analytic function. From the analyticity it follows immediately that a non-trivial bandlimited function cannot vanish on any (non-degenerate) subset of $\mathbb{S}^2$. The only function that is both bandlimited and spacelimited is the zero function.

Numerical analysis would like to deal with spacelimited functions. However, such a function (signal) of finite (space) support turns out to be non-bandlimited, so that it must contain spherical harmonics of arbitrarily large frequencies. Thus, there is a dilemma or uncertainty principle, in seeking functions that are somehow concentrated in both space and frequency.

Functions cannot have a finite support in spatial as well as in spectral domain. A certain way out is the bandlimited context of (spherical harmonics based) Shannon kernels that allows a spatiospectral concentration in terms of Slepian functions (cf. [50, 51], and the references therein), where the measure of concentration is invariably a quadratic energy ratio adapted to the local area under consideration.

Uncertainty Principle There is a way of mathematically expressing the impossibility of simultaneous confinement of a function to space and frequency (more accurately, angular momentum), namely, the *uncertainty principle* (see, e.g., [24, 49, 51]). If we consider $|F(\xi)|^2, \xi \in \mathbb{S}^2$, as a density in space so that $\|F\|^2_{L^2(\mathbb{S}^2)} = 1$, the average space (expectation) can be defined in the usual way any average is understood:

$$g_F^{\text{space}} = \int_{\mathbb{S}^2} \xi \, (F(\xi))^2 \, dS(\xi). \tag{2.11}$$

The reason for introducing an average is that it may give a gross characterization of the density. Moreover, it may indicate where the density is concentrated around the average. Various measures can be used to make certain whether the density is concentrated around the average. The most common measure is the standard deviation, σ_F^{space}, given by

$$(\sigma_F^{\text{space}})^2 = \int_{\mathbb{S}^2} (\xi - (g_F^{\text{space}}))(F(\xi))^2 \, dS(\xi). \tag{2.12}$$

The standard deviation is an indication of the space localization of the signal. If the standard deviation is small, then most of the signal is concentrated around the average space.

If $(F^{\wedge}(n, j))^2$ represents the density in frequency, then we may use it to calculate averages, the motivation being the same as in space domain. It also gives a rough idea of the main characteristics of the spectral density. The average frequency (frequency expectation), $g_F^{\text{frequency}}$, and its standard deviation, $\sigma_F^{\text{frequency}}$, (sometimes also called bandwidth) are given by

$$g_F^{\text{frequency}} = \sum_{n=0}^{\infty} \sum_{k=-n}^{n} n(n+1) \left(F^{\wedge}(n, k)\right)^2 \tag{2.13}$$

and

$$(\sigma_F^{\text{frequency}})^2 = \sum_{n=0}^{\infty} \sum_{k=-n}^{n} \left(n(n+1) - g_F^{\text{frequency}}\right)^2 \left(F^{\wedge}(n, k)\right)^2. \tag{2.14}$$

Note that, for reasons of consequence with the theory of spherical harmonics we chose $n(n+1)$ instead of the 1D-standard choice n (see, e.g., [14]).

The discovery of the uncertainty by W. Heisenberg (1927) is one of the great achievements of the last century. For signal analysis it roughly states the fact that a narrow spatial "waveform" implies a wide frequency spectrum, and a wide "spatial waveform" yields a narrow spectrum. Both the spatial waveform and the frequency spectrum cannot be made arbitrarily small simultaneously.

Space localization is at the cost of frequency localization, and vice versa. Expressed in formulas it means that

$$\Delta_F^{\text{space}} \Delta_F^{\text{frequency}} \geq 1, \tag{2.15}$$

where the so-called uncertainties Δ_F^{space}, $\Delta_F^{\text{frequency}}$ are given by

$$\Delta_F^{\text{space}} = \frac{\sigma_F^{\text{space}}}{|g_F^{\text{space}}|} \tag{2.16}$$

and

$$\Delta_F^{\text{frequency}} = (g_F^{\text{frequency}})^{1/2}. \tag{2.17}$$

The uncertainty principle enables us to understand the *transition from the theory of spherical harmonics through zonal kernel functions to the Dirac kernel*. To this end we have to realize the relative advantages of the classical Fourier expansion method by means of spherical harmonics, and this is not only in the frequency domain, but also in the space domain. It is a characteristic for Fourier techniques that the spherical harmonics as polynomial trial functions admit no localization in space domain, while in the frequency domain, they always correspond to exactly one degree, i.e., frequency, and, therefore, are said to show ideal frequency localization. Because of the ideal frequency localization and the simultaneous absence of space localization local changes of fields (signals) in the space domain affect the whole table of orthogonal (Fourier) coefficients. This, in turn, causes global changes of the corresponding (truncated) Fourier series in the space domain. Nevertheless, ideal frequency localization is often helpful for meaningful physical interpretations by relating the different observables of a geopotential to each other at a fixed frequency, e.g., the Meissl scheme in physical geodesy (see [101, 107, 108], and the references therein).

Taking these aspects on spherical harmonics modeling by Fourier series into account, trial functions which simultaneously show ideal frequency localization as well as ideal space localization would be a desirable choice. In fact, such an ideal system of trial functions would admit models of highest spatial resolution which were expressible in terms of single frequencies. However, from the uncertainty principle, space and frequency localization are mutually exclusive (Table 2.2).

In conclusion, Fourier expansion methods are well suited to resolve low- and medium-frequency phenomena, i.e., the "trends" of a signal, while their application to obtain high resolution in global or local models is critical. This difficulty is also well known to theoretical physics, e.g., when describing monochromatic

Table 2.2 Uncertainty principle and its consequences for space-frequency localization (cf. [24, 49, 51])

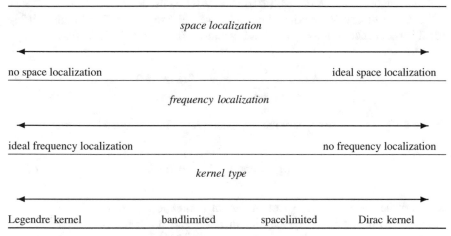

	space localization	
no space localization		ideal space localization

	frequency localization	
ideal frequency localization		no frequency localization

	kernel type		
Legendre kernel	bandlimited	spacelimited	Dirac kernel

electromagnetic waves or considering the quantum-mechanical treatment of free particles. Here, plane waves with fixed frequencies (ideal frequency localization and no space localization) are the solutions of the corresponding differential equations, but they do certainly not reflect the physical reality. As a remedy, plane waves of different frequencies are superposed into "wave-packages" that gain a certain amount of space localization while losing their ideal spectral localization. In a similar way, we are confronted with the following situation: A suitable superposition of spherical harmonics (cf. Fig. 2.10) leads to kernel functions with a reduced frequency, but increased space localization.

Zonal Kernels Any kernel function $K : \mathbb{S}^2 \times \mathbb{S}^2 \to \mathbb{R}$ that is characterized by the property that there exists a function $\tilde{K} : [0, 2] \to \mathbb{R}$ such that

$$K(\xi, \eta) = \tilde{K}(|\xi - \eta|) = \tilde{K}\left(\sqrt{2 - 2\xi \cdot \eta}\right) = \hat{K}(\xi \cdot \eta), \quad \xi, \eta \in \mathbb{S}^2, \qquad (2.18)$$

is called a (spherical) *radial basis function* (or *zonal function*, at least in the theory of constructive approximation). The application of a rotation (i.e., a 3×3 orthogonal matrix $\mathbf{t}$ with $\mathbf{t}^T = \mathbf{t}^{-1}$) leads to

$$K(\mathbf{t}\xi, \mathbf{t}\eta) = \hat{K}((\mathbf{t}\xi) \cdot (\mathbf{t}\eta)) = \hat{K}(\xi \cdot (\mathbf{t}^T \mathbf{t}\eta)) = \hat{K}(\xi \cdot \eta) = K(\xi, \eta). \qquad (2.19)$$

In particular, a rotation around the axis $\xi \in \mathbb{S}^2$ (i.e., we have $\mathbf{t}\xi = \xi$) yields $K(\xi, \eta) = K(\xi, \mathbf{t}\eta)$ for all $\eta \in \mathbb{S}^2$. Hence, $K(\xi, \cdot)$ possesses rotational symmetry with respect to the axis ξ.

A kernel $\hat{K} : \mathbb{S}^2 \times \mathbb{S}^2 \to \mathbb{R}$ satisfying $\hat{K}(\xi \cdot \eta) = \hat{K}(\mathbf{t}\xi \cdot \mathbf{t}\eta), \xi, \eta \in \mathbb{S}^2$, for all orthogonal transformations $\mathbf{t}$ is known as a zonal kernel function.

To highlight the reducibility of $\hat{K}$ to a function defined on the interval $[-1, 1]$, the notation $(\xi, \eta) \mapsto \hat{K}(\xi \cdot \eta)$, $(\xi, \eta) \in \mathbb{S}^2 \times \mathbb{S}^2$, is used.

From the theory of spherical harmonics we get a representation of any $L^2(\mathbb{S}^2)$-zonal kernel function K in terms of a Legendre expansion (cf. Fig. 2.10)

$$K(\xi \cdot) = \sum_{n=0}^{\infty} \frac{2n+1}{4\pi} K^{\wedge}(n) P_n(\xi \cdot) \tag{2.20}$$

(in the $\| \cdot \|_{L^2(\mathbb{S}^2)}$-sense), where the sequence $\{K^{\wedge}(n)\}_{n \in \mathbb{N}_0}$ given by

$$K^{\wedge}(n) = 2\pi \int_{-1}^{1} K(t) P_n(t) \, dt \tag{2.21}$$

is called the *Legendre symbol* of the zonal kernel $K(\xi \cdot)$. A simple but extreme example (with optimal frequency localization and no space localization) is the Legendre kernel, where $K^{\wedge}(n) = 1$ for one particular n and $K^{\wedge}(m) = 0$ for $m \neq n$, i.e., the Legendre kernel is given by

$$\mathbb{S}^2 \times \mathbb{S}^2 \ni (\xi, \eta) \mapsto \frac{2n+1}{4\pi} P_n(\xi \cdot \eta). \tag{2.22}$$

Additive clustering of weighted Legendre kernels generates zonal kernel functions. We distinguish bandlimited kernels (i.e., $K^{\wedge}(n) = 0$ for all $n > N$) and non-bandlimited ones, for which infinitely many numbers $K^{\wedge}(n)$ do not vanish.

Non-bandlimited kernels show a much stronger space localization than their bandlimited counterparts. Empirically, if

$$K^{\wedge}(n) \approx K^{\wedge}(n+1) \approx 1 \tag{2.23}$$

for many successive large integers n, then the support of the series (2.20) in the space domain is small, i.e., the kernel is spacelimited (i.e., in the jargon of approximation theory "locally supported"). This leads to the other extremal kernel (in contrast to the Legendre kernel) which is the Dirac kernel with optimal space localization but no frequency localization and $K^{\wedge}(n) = 1$ for all n. However, the Dirac kernel does not exist as a classical function in mathematical sense, it is a generalized function (i.e., distribution). Nevertheless, it is well known that, if we have a family of kernels $\{K_J\}_{J=0,1,\ldots}$, where $\lim_{J \to \infty} K_J^{\wedge}(n) = 1$ for each n and an additional (technical) condition holds, then $\{K_J\}_{J=0,1,\ldots}$ is an *approximate identity*, i.e., $K_J * F$ tends to F in the sense of $L^2(\mathbb{S}^2)$ for all $F \in L^2(\mathbb{S}^2)$, or to $F \in C^{(0)}(\mathbb{S}^2)$ for all $F \in C^{(0)}(\mathbb{S}^2)$, respectively.

Assuming $\lim_{n \to \infty} K^{\wedge}(n) = 0$ we are led to the assertion that the slower the sequence $\{K^{\wedge}(n)\}_{n=0,1,\ldots}$ converges to zero, the lower is the frequency localization, and the higher is the space localization. A unifying scheme can be found in Table 2.3.

Table 2.3 Different types of zonal kernels: bandlimited, spacelimited, and non-spacelimited/non-bandlimited (cf. [46, 51])

General case			
Legendre kernels			Dirac kernel
$K^\wedge(n) = \delta_{n,k}$	Bandlimited	Spacelimited	$K^\wedge(n) = 1, n = 0, \ldots$
	$K^\wedge(n) = 0,\ n > N$	$K(\xi \cdot \eta) = 0, 1 - \xi \cdot \eta \geq \delta$	
	Shannon	Haar	
	$K^\wedge(n) = 1,\ n \leq N$	$K(\xi \cdot \eta) = 1, 1 - \xi \cdot \eta \leq \delta$	

Zonal kernel function theory relies on the following principles:

1. Weighted Legendre kernels are the summands of zonal kernel functions.
2. The Legendre kernel is ideally localized in frequency. The Dirac kernel is ideally localized in space.
3. The only frequency and spacelimited zonal kernel is the zero function.

All in all, kernels exist as bandlimited and non-bandlimited functions (see Table 2.3). Every bandlimited kernel refers to a cluster of a finite number of polynomials, i.e., spherical harmonics; hence, it corresponds to a certain band of frequencies. In contrast to a single polynomial which is localized in frequency but not in space, a bandlimited kernel such as the Shannon kernel already shows a certain amount of space localization. If we move from bandlimited to non-bandlimited kernels the frequency localization decreases and the space localization increases in accordance with the relationship provided by the uncertainty principle. Kernel function approximation exists in spline and wavelets specifications naturally based on certain realizations in frequency and space localization. Obviously, if a certain accuracy should be guaranteed in kernel function approximation, adaptive sample point grids are required for the resulting extension of the spatial area determined by the kernels under investigation.

Constructive Approximation Methods If data sets on the sphere are localized in size, typically by spherical rectangles, approximation problems can be attacked by application of *tensor product techniques* in terms of polar coordinates originally designed for bivariate Euclidean space nomenclature. However, global problems like the determination of the gravitational field, magnetic field, tectonic movements, ocean circulation, climate change, hydrological and meteorological purposes, etc., involve essentially the entire surface of the sphere, so that modeling the data as arising in Euclidean two-space via latitude–longitude coordinate separation is no longer appropriate. Even more, since there is no differentiable mapping of the entire sphere to a bounded planar region, there is a need to develop approximations such as sampling methods over the sphere itself, thereby avoiding (artificially occurring) singularities. Looking at a numerically efficient and stable global model in today's literature, a spherical (and usually not a physically more suitable ellipsoidal) reference shape of the Earth has been taken into account in almost all practical approximations because of their conceptual simplicity and numerical computability.

Starting from the time of Laplace, Legendre, and Gauss (see, e.g., [17, 53, 72]), the context of spherical harmonics is a well-established tool, particularly for access to the inversion of problems under the assumption of a bandlimited Earth's gravitational and/or magnetic model. Nowadays, spectral reference models, i.e., *Fourier expansions in terms of spherical harmonics* for the Earth's gravitational and magnetic potential, are widely known by tables of expansion coefficients as frequency determined constituents. In this respect, it should be mentioned that geoscientific modeling demands its own nature. Concerning modeling one is usually not interested in the separation into scalar Cartesian component functions involving product ingredients. Instead, inherent physical properties should be observed. For example, the deflections of the vertical form a vector-isotropic surface gradient field on the Earth's surface, the equations for (geostrophic) ocean (surface) flow involving geoidal undulations (heights) imply a divergence-free vector-isotropic nature, and satellite gradiometer data lead to tensor-isotropic Hesse fields. As a consequence, in a geoscientifically reflected spherical framework, all these physical constraints result in a formulation by rotation-invariant pseudodifferential operators. Hence, rotational symmetry (isotropy) is an indispensable ingredient (see, e.g., [49, 115]) in the bridging transformation that relates geophysical and/or geodetic quantities, i.e., the object parameters, to the observed and/or measured data sets, and vice versa.

Splines Commonly, zonal functions on the sphere recognized as positive definite kernels may be interpreted as generating reproducing kernels of Sobolev spaces. This is the reason why *spherical splines* may be based on a variational approach (cf. [21–23, 49–51, 119, 120]) that minimizes a weighted Sobolev norm of the interpolant, with a large class of spline manifestations provided by pseudodifferential operators being at the disposal of the user.

Sobolev space framework involving rotation-invariant pseudodifferential operators (as originated by observables, e.g., in physical geodesy) shows some important benefits of spline interpolation as preparatory tool for spherical sampling. Accordingly we are confronted with the following situation:

1. Interpolating/smoothing splines are canonically based on a variational approach that minimizes a weighted Sobolev norm of the interpolating/ smoothing spline, with a large class of spline manifestations provided by pseudodifferential operators being at the disposal of the user. Regularly distributed as well as scattered data systems can be handled.
2. Artificial singularities caused by the use of (polar) coordinates in global approximation can be avoided totally.
3. The rotational invariance of observables (such as gravity anomalies, gravity disturbances, and disturbing potentials in geodetic theory) is perfectly maintained.
4. Measurement errors can be handled by an adapted interpolation/smoothing spline method. Error bounds can be derived that include computable constants rather than only being given in terms of the order of convergence of the maximum distance from any point in the domain to the nearest data point.
5. Spline spaces serve as canonical reference spaces for the purposes of spherical sampling relative to finite as well as infinite scattered data distributions.

6. Spherical splines provide approximations using kernel functions with a fixed "window," i.e., preassigned frequency and space relation.
7. The accuracy of spline approximation can be controlled by a decreasing sampling width.
8. Global spherical spline interpolation in terms of zonal, i.e., radial basis functions, has its roots in physically motivated problems of minimizing a "(linearized) curvature energy" variational problem consistent to data points. Numerical experiences with the linear system of equations have shown that the system tends to be ill-conditioned unless the number of data points is not too large. Clearly, oscillation phenomena in spline interpolation may occur for larger data gaps.

The extension of spherical splines to harmonic splines started with [22] and [110]. Further developments and results on harmonic splines can be found in [25, 30, 35, 50].

Wavelets The wavelet approach is a more flexible approximation method than the Fourier expansion approach in terms of spherical harmonics or the variational spline method using zonal kernels with fixed window. Due to the fact that variable "window kernel functions," i.e., zooming-in, are being applied, a substantially better representation of the high-frequency "short-wavelength" part of a function is achievable (global to scale-dependent local approximation). The zooming-in procedure allows higher global resolutions and, therefore, makes a better exposure of the strong correlation between the function (signal) under consideration and the local phenomena that should be modeled. Furthermore, the multiscale analysis can be used to modify and improve the standard approach in the sense that a local approximation can be established within a global orthogonal (Fourier) series and/or spline concept (see [49]).

In essence, the characterization of *spherical wavelets* contains three basic attributes:

1. *Basis Property.* Wavelets are building blocks for the approximation of arbitrary functions (signals). In mathematical terms this formulation expresses that the set of wavelets forms a "frame" (see, e.g., [15] and the references therein for details in classical one-dimensional theory).
2. *Decorrelation.* Wavelets possess the ability to decorrelate the signal. This means that the representation of the signal via wavelet coefficients occurs in a "more constituting" form than in the original form, reflecting any certain amount of space and frequency information. The decorrelation enables the extraction of specific information contained in a signal through a particular locally reflected number of coefficients. Signals usually show a correlation in the frequency domain as well as in the space domain. Obviously, since data points in a local neighborhood are more strongly correlated as those data points far-off from each other, signal characteristics often appear in certain frequency bands. In order to analyze and reconstruct such signals, we need "auxiliary functions" providing localized information in the space as well as in the frequency domain. Within a "zooming-in process," the amount of frequency as well as space contribution can be specified in a quantitative way. A so-called scaling function

forms a compromise in which a certain balanced amount of frequency and space localization in the sense of the uncertainty principle is realized. As a consequence, each scaling function on the sphere depends on two variables, namely, a rotational parameter (defining the position) and a dilational (scaling) parameter, which controls the amount of the space localization to be available at the price of the frequency localization, and vice versa. Associated with each scaling function is a wavelet function, which in its simplest manifestation may be understood as the difference of two successive scaling functions. Filtering (convolution) with a scaling function takes the part of a low-pass filter, while convolution with the corresponding wavelet function provides a band-pass filtering. A multiscale approximation of a signal is the successive realization of an efficient (approximate identity realizing) evaluation process by use of scaling and wavelet functions which show more and more space localization at the cost of frequency localization. The wavelet transform within a multiscale approximation lays the foundation for the decorrelation of a signal.

3. *Efficient Algorithms.* Wavelet transform provides efficient algorithms because of the space localizing character. The successive decomposition of the signal by use of wavelets at different scales offers the advantage for efficient and economic numerical calculation. The detailed information stored in the wavelet coefficients leads to a reconstruction from a rough to a fine resolution and to a decomposition from a fine to a rough resolution in the form of tree algorithms. In particular, the decomposition algorithm is an excellent tool for the postprocessing of a signal into "constituting blocks" by decorrelation, e.g., the specification of signature bands corresponding to certain signal specifics.

As a consequence, spherical wavelets may be regarded as constituting multiscale building blocks, which provide a fast and efficient way to decorrelate a given signal data set. The properties (basis property, decorrelation, and efficient algorithms) are common features of all wavelets, so that these attributes form the key for a variety of applications particularly for signal reconstruction and decomposition, thresholding, data compression, denoising by use of, e.g., multiscale signal-to-noise ratio, etc. The essential power of spherical wavelets is based on the "zooming-in" property, i.e., scale-dependent varying amounts of both frequency and space localization. This multiscale structure is the reason why spherical wavelets can be used as mathematical means for breaking up a complicated structure of a function into many simple pieces at different scales and positions. It should be pointed out that several wavelet approaches involving spherical wavelets have been established, all of them providing multiscale approximation, but not all of them showing a structural "breakthrough" in the form of a multiresolution of the whole reference space by nested detail spaces. In all cases, however, wavelet modeling is provided by a two-parameter family reflecting the different levels of localization and scale resolution.

Often in geosciences fast approximation procedures are required for large amounts of data. Whereas global methods like approximation with spherical harmonics have proven to be reliable for global trend resolution, the focus for local reconstructions has shifted to "zooming-in" techniques involving wavelets. It is evident that wavelet approximation on (parts of) $\mathbb{R}^2$ is much better studied (see, e.g., [75]). In fact, there exists a huge number of wavelets for very different purposes. In particular, there is a variety of wavelets which are orthogonal or show other very interesting properties such as a compact support (cf. [15]). Specifically, the usage of filter banks makes a 2D-wavelet transform very economical. A disadvantage is that one cannot easily deal with data which are not given on a regular grid. Nonetheless there are, e.g., lifting schemes for scattered data (cf. [70]), so that the speed disadvantage in comparison to grid based methods is manageable. Another possibility of a plane reduction is the use of non-grid based FFT techniques [105] and the realization of wavelet transform via the known representation in the 2D-Fourier domain. At this point, it should be noted that intrinsic situations on the sphere are rather different. All wavelet methods usually suffer from the property that they are not canonically parallelizable. Usually, the characteristic feature of wavelets is that a single function, the so-called mother-wavelet (see, e.g., [15]), is dilated to cover different frequency bands and is shifted to cover the spatial domain. Any function possessing a cap as a support cannot be shifted around on the sphere in order to cover it without overlap. This observation considerably limits the use of an isotropic mother-wavelet because one has to solve larger systems of equations to counter the non-orthogonality of the wavelets.

Roughly speaking, the wavelet transform is a space localized replacement of the Fourier transform, providing space-varying frequency distribution in banded form. Wavelets provide sampling by only using a small fraction of the original information of a function. Typically the decorrelation is achieved by wavelets which have a characteristic local band (localization in frequency). Different types of wavelets can be established from certain constructions of space/frequency localization. It turns out that the decay towards long and short wavelengths (i.e., in information theoretic jargon and band-pass filtering) can be assured without any difficulty. Moreover, vanishing moments of wavelets enable us to combine polynomial (orthogonal) expansion (responsible for the long-wavelength part of a function) with wavelet expansions responsible for the medium-to-short-wavelength contributions. Because of the rotation-invariant nature of a large number of geodetic and geophysical quantities, resulting sampling methods of zonal spline and wavelet nature have much to offer. This is the reason why the authors decided to add to the plane involved matter some of the significant sphere intrinsic features and results of spherical sampling in a unifying concept in the second half of the book. Furthermore, the fundamental idea of handling inverse problems of satellite technology (such as satellite-to-satellite tracking and satellite gravity gradiometry) is to understand regularization of inverse (non-bounded) pseudodifferential operators by a multiresolution analysis using certain kernel function expressions as regularizing wavelets.

It is also worth mentioning that future spaceborne observation combined with terrestrial and airborne activities will provide huge data sets of the order of millions of data to be continued downward to the Earth's surface. Standard mathematical theory and numerical methods are not at all adequate for the solution of data systems with a structure such as this, because these methods are simply not adapted to the specific character of the spaceborne problems. They quickly reach their capacity limit even on very powerful computers. In our opinion, a reconstruction of significant geophysical quantities from future data material requires much more: for example, it requires a careful analysis, fast solution techniques, and a proper stabilization of the solution, usually including procedures of regularization (see [25] and the references therein). In order to achieve these objectives various strategies and structures must be introduced reflecting different aspects. As already pointed out, while global long-wavelength modeling can be adequately done by the use of polynomial expansions, it becomes more and more obvious that splines and/or wavelets are most likely the candidates for medium- and short-wavelength approximation. But the multiscale concept of wavelets demands its own nature which—in most cases—cannot be developed from the well-known theory in Euclidean spaces. In consequence, the stage is also set to present the essential ideas and results involving a multiscale framework to the geoscientific community.

All in all, wavelets are able to find a possibility to break up complicated functions like the geomagnetic field, electric currents, gravitational field, deformation field, oceanic currents, etc., into single pieces of different resolutions and to analyze these pieces separately. This helps to find adaptive methods (cf. Figs. 2.11, 2.12) that take into account the specific structure of the data, i.e., in areas where the data show only a few coarse spatial structures the resolution of the model can be chosen to be rather low; in areas of complicated data structures the resolution can be increased accordingly. In areas where the accuracy inherent in the measurements is reached the solution process can be stopped by some kind of thresholding. That is, using scaling functions and wavelets at different scales, the corresponding approximation techniques can be constructed as to be suitable for the particular data situation. Consequently, although most data show correlation in space as well as in frequency, the kernel functions with their simultaneous space and frequency localization allow for the efficient detection and approximation of essential features in the data structure by only using fractions of the original information (decorrelation of signatures).

In geosciences we usually consider a separable Hilbert space such as the L^2-space (of functions having finite signature energy) with a (known) polynomial basis as reference space for *ansatz functions*. However, there is a striking difference between the L^2-space over the Earth's body and the Earth's surface. Continuous "surface functions" can be described in arbitrary accuracy, for example, with respect to $C^{(0)}$- and L^2-topology by restrictions of harmonic functions (such as homogeneous harmonic polynomials), whereas continuous "volume functions" contain anharmonic ingredients (for more details see, e.g., [30, 35, 40, 82, 84]). This fact has serious consequences for the reconstruction of signatures on balls in $\mathbb{R}^3$ which are currently under investigation, e.g., by Michel (cf. [83]) and his members of the Geomathematics Group Siegen.

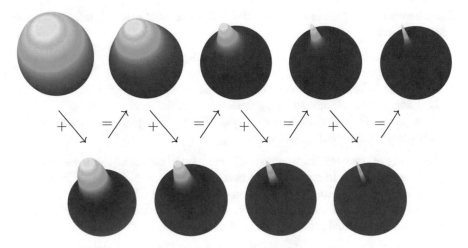

Fig. 2.11 Scaling functions (upper row) and wavelet functions (lower row) in mutual relation ("tree structure") within a multiscale approximation (cf. [27, 51])

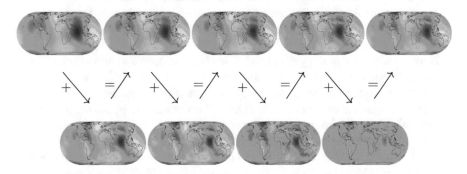

Fig. 2.12 Global Earth's Gravitational Model (EGM) (cf. [27, 51, 102]): Scaling function (upper row) and wavelet function (lower row) reconstruction in mutual relation (i.e., "tree structure"). On the one hand, the multiscale approximation of EGM shows that the gravitational potential is a rather smooth function, so that most of its structure can be modeled by low-pass filtering. On the other hand, EGM is sharply decorrelated by band-pass filtering, so that the detailed structure for larger scales concentrates on geophysically relevant zones (e.g., subduction zones, areas of strong orogenese, etc.)

2.6   Geomathematics as General Methodology

Current methods of applied measurement and evaluation processes vary strongly, depending on the measurement parameters (gravity, electric or magnetic field force, temperature and heat flow, etc.), the observed frequency domain, and the occurring basic "field characteristic" (potential field, diffusion field, or wave field, depending on the basic differential equations). In particular, the differential equation strongly influences the evaluation processes. The typical mathematical methods are therefore listed here according to the respective "field characteristic"—as it is usually done in geomathematics.

- *Potential methods* (potential fields, elliptic differential equations) in geomagnetics, geoelectrics, gravimetry, geothermal research, etc.
- *Diffusion methods* (diffusion fields, parabolic differential equations) in flow and heat transport, magnetotellurics, geoelectromagnetics, etc.
- *Wave methods* (wave fields, hyperbolic differential equations) in exploration seismology, georadar measurements, etc.

The diversity of mathematical methods will increase in the future due to new technological developments in computer and measurement technology. More intensively than before, we must aim for the creation of models and simulations for combinations and networks of data and observable structures. The mathematical process for the solution of practical problems usually has the following components:

- *"Transfer" from Applications to the Language of Mathematics*: The practical problem is translated into the language of mathematics, requiring the cooperation between application-oriented scientists and mathematicians.
- *Mathematical Analysis*: The resulting mathematical problem is examined as to its "well-posedness" (i.e., existence, uniqueness, and dependence on the input data).
- *Development of a Mathematical Solution Method*: Appropriate analytical, algebraic, and/or numerical methods and processes for a specific solution must be adapted to the problem; if necessary, new methods must be developed. The solution process is carried out efficiently and economically by the decomposition into individual operations, usually on computers.
- *"Back-Transfer" from the Language of Mathematics to Applications*: The results are illustrated adequately in order to ensure their evaluation. The mathematical model is validated on the basis of real data and modified, if necessary. We aim for good accordance of model and reality.

Often, the process must be applied several times in an iterative way in order to get a sufficient insight into the system Earth. Nonetheless, the advantage and benefit of the mathematical processes are a better, faster, cheaper, and more secure problem solution on the basis of the mentioned means of simulation, visualization, and reduction of large amounts of data. So, what is it exactly that enables mathematicians to build a bridge between the different disciplines? As already pointed out, the mathematics' world of numbers and shapes contains very efficient tokens by which we can describe the rule-like aspect of real problems. This description includes a simplification by abstraction: essential properties of a problem are separated from unimportant ones and included into a solution scheme. Their "eye for similarities" often enables mathematicians to recognize a posteriori that an adequately reduced problem may also arise from very different situations, so that the resulting solutions may be applicable to multiple cases after an appropriate adaptation or concretization. Without this second step, abstraction remains essentially useless.

The interaction between abstraction and concretization characterizes the history of mathematics and its current rapid development as a common language and independent science. A problem reduced by abstraction is considered as a new "concrete" problem to be solved within a general framework, which determines the validity of a possible solution. The more examples one knows, the more one recognizes the causality between the abstractness of mathematical concepts and their impact and cross-sectional importance.

Chapter 3
Exemplary Applications: Novel Exploration Methods

Next we present two novel exploration methods, thereby using the structure of a methodological circuit (as presented in Sect. 2.6), respectively. We start with inverse gravimetry, which becomes an increasing importance, e.g., in geothermal research. Then we go over to a standard technique in geoexploration, namely reflection seismics, for which a "mollifier inversion procedure" similar to the approach in gravimetry will be developed.

3.1 Circuit: Gravimetry

An essential subsystem in Earth's system research is the gravity field of the Earth.

The *gravity acceleration* (gravity) w is the resultant of the *gravitation v* and the *centrifugal acceleration c*, such that $w = v + c$ (cf. Fig. 3.1). The centrifugal force arises as a result of the rotation of the Earth about its axis. So, in terrestrial observation, if we are able to eliminate the centrifugal influence from gravity, we end up with gravitation on the Earth's surface.

According to the classical Newtonian approach this fact enables us to deal with the *problem of inverse gravimetry*.

> The problem of inverse gravimetry is to derive the density distribution inside the Earth from an appropriate knowledge of (measurable quantities of the) Earth's gravitation for a certain subset of the Earth's crust.

In what follows, we first give some background information of the Earth's gravitational field and its measurement by gravimetry (see also [30]). Second, we present a novel multiscale technique for modeling density contrasts inside the Earth's crust. Our main interest is to develop a geomathematical method to specify certain geological features, e.g., by a decorrelation process using geophysically oriented wavelet structures. The philosophy behind our considerations is that every

W. Freeden et al., *An Invitation to Geomathematics*, Lecture Notes in Geosystems Mathematics and Computing, https://doi.org/10.1007/978-3-030-13054-1_3

Fig. 3.1 Gravitation v,
centrifugal acceleration c, and
gravity acceleration w

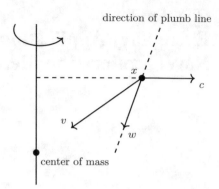

geological formation can be found in a particular wavelet band inside the density signatures and, under some circumstances, it is also detectable in a corresponding potential band obtainable from the input measurements.

The problem of gravimetry is that we have to deal with an ill-posed inverse problem, which violates all Hadamard's requirements of existence, uniqueness, and stability (cf. [41]). Nevertheless, when applied to a representative test model, the presented regularization method by "mollification" shows acceptable geological results up to a certain depth, and the multiscale modeling becomes better and better the more information inside the Earth's is known, for example, from in-borehole gravimetry data available from geothermic power plant projects (note that, for simplicity, our approach is restricted to evaluation functionals of the Newtonian gravitational potential, which is unrealistic until now. For practical purposes, other geodetic observables have to come into play such as gravity anomalies, gravity disturbances, and deflections of the vertical. However, mathematically those gravitational observables involve certain derivatives applied to the gravitational potential, i.e., the Newton volume integral, so that a multiscale approach can be handled in quite similar way).

Background Knowledge It must be emphasized that it is not the objective of our study to go into the nature of gravitation. This would be a subject of physics and would require a study of Newton's and Einstein's theories of gravitation and more recent developments in this field. We would rather like to get a feeling on how observations of the gravitational field, more specifically gravimetric measurements, can be taken and how mathematical use of them can be made for purposes of exploration (Fig. 3.2).

The force of the gravity provides a directional structure. It is tangential to the vertical plumb lines and perpendicular to all (level) equipotential surfaces. Any water surface at rest is part of a level surface. Level (equipotential) surfaces are ideal reference surfaces, for example, for heights. As the level surfaces are, so to speak, "horizontal," they play an important part in our daily life (e.g., in civil engineering for the purpose of height determination). Due to Listing [73], the geoid is defined as that level surface of the gravity field which best fits the mean sea level (for more detailed geodetic information see, e.g., [52, 54, 57, 64, 65, 67, 79, 81–83, 86, 88, 106, 117, 118]).

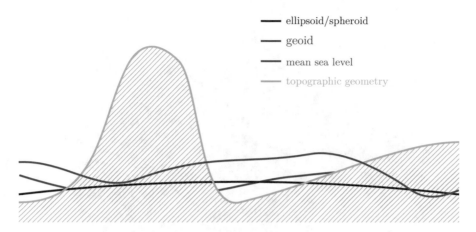

Fig. 3.2 Geodetically relevant geometries (from [47])

Gravity Field and Important Applicabilities The gravity field is of great importance for many applications (see Fig. 3.3) from which we only mention five significant aspects (cf. [106]):

- *Geodesy and Civil Engineering.* Accurate heights are needed for civil constructions, mapping, etc. They are obtained by leveling, a very time-consuming and expensive procedure. Today's geometric heights can be obtained fast and efficiently from space positioning (for example, the *Global Navigation Satellite System*, GNSS). The geometric heights are convertible to leveled heights by subtracting the precise geoid, which is implied by a high resolution gravitational potential. To be more specific, in those areas where good gravity information is already available, the future data situation will eliminate all medium and long-wavelength distortions in unsurveyed areas. GNSS (i.e., NAVSTAR GPS, USA; GLONASS, Russian Federation; GALILEO, EU; BEIDOU, People's Republic of China) together with the planned explorer satellite missions will provide high quality height information at global scale.
- *Solid Earth Physics.* The gravity anomaly field has its origin mainly in mass inhomogeneities of the continental and oceanic lithosphere. Together with height information and regional tomography, a much deeper understanding of tectonic processes should be obtainable.
- *Physical Oceanography.* Altimeter (radar) satellites in combination with a precise geoid will deliver global dynamic ocean topography. From the global surface circulation and its variations in time, the ocean topography can be computed resulting in a completely new dimension of accuracy. Surface circulation allows the determination of transport processes of, e.g., polluted material.
- *Earth Climate System.* There is a growing awareness of global environmental problems (for example, the CO_2-question, the rapid decrease of rain forests, global sea level changes, etc.). What is the role of the future airborne methods and satellite missions in this context? They do not tell us the reasons for physical

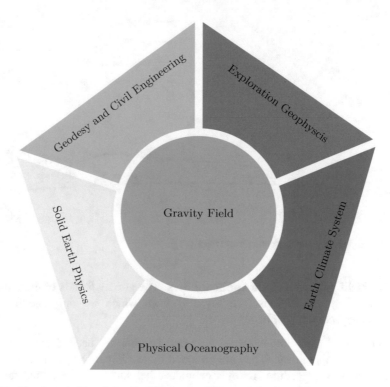

Fig. 3.3 The knowledge of the Earth's gravity field and its essential importance for many applications (from [47])

processes, but it is essential to bring the phenomena into one system (e.g., to make sea level records comparable in different parts of the world). In other words, the geoid is viewed as an almost static reference for many rapidly changing processes and at the same time as a "frozen picture" of tectonic processes that evolved over geological time spans.

- *Exploration.* Strong improvements of GNSS-based data sets can be expected from the future scenario. Spaceborne gravity, of course, has a great advantage because measurements of the gravity field are not restricted to certain areas. For local purposes of exploration, however, today's GNSS-based gravitational data sets are until now only helpful for an a priori trend modeling, particularly for boundary strips of a local test area, where numerical instabilities have to be expected. In fact, gravity anomalies of dimension very much greater than the gravity anomalies caused by, e.g., oil, gas, and water structures are of little significance for exploration. The fundamental interest in gravitational methods in exploration today therefore is still based on terrestrial measurements of small variations, i.e., terrestrial gravimetry.

Table 3.1 Scientific use of gravitational field observables

Solid Earth	Oceanography	Glaciology	Geodesy	Climate
Crustal density	Dynamic topography	Bedrock topography	Leveling (GNSS)	Sea level changes
Post glacial rebound	Heat transport	Flux	Height systems	Coastal zones
	Mass transport		Orbit determination	

Finally, we give a list of scientific disciplines (cf. Table 3.1), where gravitational field observables play a major role:

From Gravitation to Potential Theory   Our interest is merely laid in studying the gravitational field in macroscopic sense, where the quantum behavior of gravitation may not be taken into account. Moreover, in geodetically reflected Earth's gravity work, velocities that are encountered are considerably smaller than the speed of the light. As a consequence, Newtonian physics can be used safely.

Gravitation   From a preparatory methodological point of view two steps should be observed:

- The first (practical) step is to gain experience and collect empirical data. Here, mathematics is used as a kind of registration or catalog to get concise descriptions of the laws of nature in empirical formulas.
- The second (theoretical) step is to set up a system of mathematical definitions and laws from which these formulas and more results may be derived by mathematical arguing.

In the sense of the first step, Isaac Newton often told the story that he was inspired to develop his theory of gravitation by watching the fall of an apple ("test mass") from tree. Although it has been said that the apple story is a myth, some acquaintances of Newton do, in fact, confirm the incident. In other words, by the registration process of a falling test mass, Newton was able to discover his famous law about the mutual attraction of two masses, telling us that the attractive force, i.e., gravitation, is directed along the line connecting the two centers of the objects and is proportional to both masses and the squared inverse of the distance between the two objects.

In the sense of the second step, the basic Newtonian concepts of gravitation (see Fig. 3.4) could be naturally extended to potential theory so as to give a complete description of the laws and relations (see, e.g., [30]):

(a) *Potential of a mass point:* According to *Newton's law of gravitation* two points x, y with masses M_x, M_y attract each other with a force given by

$$-\frac{\gamma}{4\pi} \frac{M_x M_y}{|x-y|^3}(x-y), \quad x, y \in \mathbb{R}^3, \, x \neq y. \tag{3.1}$$

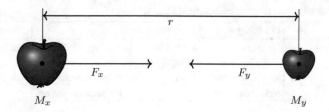

Fig. 3.4 Gravitation between point masses, i.e., two mass points with a distance $r = |x-y|$ attract each other by a force with moduli fulfilling $F_x = F_y = \gamma\, M_x M_y / r^2$, where γ is the gravitational constant (cf. [47])

The force is directed along the line connecting the two points x, y. The constant γ denotes Newton's gravitational constant (note that γ can be assumed to be equal to one in the theoretical part, but not in numerical computations).

Although the masses M_x, M_y attract each other in a symmetric way, it is convenient to call one of them the *attracting mass* and the other one the *attracted mass*. Conventionally the attracted mass is set equal to unity and the attracting mass is denoted by $M = M_y$:

$$v(x) = -\frac{\gamma}{4\pi}\,\frac{M}{|x-y|^3}(x-y), \quad x \in \mathbb{R}^3 \backslash \{y\}. \tag{3.2}$$

The formula (3.2) expresses the force exerted by the mass M on a unit mass located at the distance $|x - y|$ from M. Obviously, the intensity $|v(x)|$ of the gravitational force $v(x)$ is given by

$$|v(x)| = \frac{\gamma}{4\pi}\,\frac{M}{|x-y|^2}, \quad x \in \mathbb{R}^3 \backslash \{y\}. \tag{3.3}$$

The scalar function V defined by $V(x) = \gamma\, M\, G(\Delta; |x-y|)$ with the so-called Newton kernel

$$G(\Delta; |x-y|) = \frac{1}{4\pi}\,\frac{1}{|x-y|}, \quad x \in \mathbb{R}^3 \backslash \{y\} \tag{3.4}$$

is called the *potential of gravitation* at the point y. The force vector $v(x)$ is the gradient vector of the scalar $V(x)$: $v(x) = \nabla V(x)$, $x \in \mathbb{R}^3 \backslash \{y\}$. Calculating the divergence $\nabla \cdot$ of the gradient field v, it readily follows that $\nabla \cdot v(x) = \nabla \cdot \nabla V(x) = \Delta V(x) = 0$, $x \in \mathbb{R}^3 \backslash \{y\}$ (note that $|y| \leq \frac{|x|}{2}$ implies $|x-y| \geq ||x| - |y|| \geq \frac{1}{2}|x|$, i.e., V is regular at infinity: $V(x) = O(|x|^{-1})$, $|x| \to \infty$).

(b) *Potential of a finite mass point system:* The potential for N points x_i with masses M_i, $i = 1, \ldots, N$, is the sum of the individual contributions

$$V(x) = \gamma \sum_{i=1}^{N} M_i\, G(\Delta; |x - y_i|), \quad x \in \mathbb{R}^3 \backslash \{y_1, \ldots, y_n\}. \tag{3.5}$$

Clearly we have $\Delta V(x) = 0$, $x \in \mathbb{R}^3 \backslash \{y_1, \ldots, y_N\}$.

(c) *Potential of a volume:* Let $\mathcal{G} \subset \mathbb{R}^3$ be a *regular region* (for a more detailed description see [30]). The point masses are distributed continuously over $\mathcal{G} \subset \mathbb{R}^3$ with density F. Then the discrete sum (3.5) becomes a continuous sum, more accurately, an integral over the body $\mathcal{G}$:

$$V(x) = \gamma \int_{\mathcal{G}} G(\Delta; |x - y|) F(y) \, dy. \tag{3.6}$$

Obviously, the potential V is harmonic in $\mathbb{R}^3 \backslash \overline{\mathcal{G}}$, i.e., V satisfies the *Laplace equation* $\Delta V(x) = 0$, $x \in \mathbb{R}^3 \backslash \overline{\mathcal{G}}$. Note that V is defined on the whole space $\mathbb{R}^3$, however, $\Delta V(x)$ may not be obtained easily by interchanging the Laplace operator Δ and the integral over $\mathcal{G}$ for all points x inside $\mathcal{G}$. As a matter of fact (see, e.g., [30]), under Hölder-continuity imposed on F, the *Poisson differential equation*

$$\nabla \cdot v(x) = \Delta V(x) = \Delta_x \gamma \int_{\mathcal{G}} G(\Delta; |x - y|) \, F(y) \, dy = -\gamma \, \alpha(x) \, F(x), \quad x \in \overline{\mathcal{G}}. \tag{3.7}$$

is valid, where $\alpha(x)$ is the solid angle at x subtended by the surface $\partial \mathcal{G}$. At infinity the potential behaves like $V(x) = O(|x|^{-1})$, $|x| \to \infty$, uniformly with respect to all directions, i.e., V is regular at infinity.

Potential theory as the scientific collection of ideas, concepts, and structures involving the Laplace operator gained new aspects when discussing certain features of the Earth's gravitational field, and it was challenged with new problems, of which the boundary value problems probably are the best known. Potential theory actually guarantees that if certain values of a potential under specific consideration are given on the boundary of a closed body, the potential is determined via the boundary value problem in the interior (or in the exterior when an additional regularity condition at infinity is supposed to hold true). This assertion, of course, has been checked in many experiments also under geodetic and geophysical auspices, but naturally it cannot be verified experimentally in the generality in which it can be stated mathematically. More specifically, at a stage, where the theory is regarded as satisfactory from the physicist's point of view, it is a system of fundamental laws, definitions, and problems, of which some, under certain conditions, have been solved mathematically. The problems in their full generality, however, are given to mathematics as conjectures, in a sense, to be proved. They become the object of a study of the well-posedness, i.e., existence, uniqueness, and stability proofs, which therefore aim at establishing the consistency of the general physical theory (cf. [63]). So, the aspects of potential theory have changed considerably when constituents could be described by means of Laplace's equation, just as scientific tasks arose from the theory of stationary flow, that indeed uses the same differential equation. It could thus be observed how new physical applications developed new aspects of potential theory and the theory of partial differential equations, primarily originated on gravitational developments.

In geopotential theory, as in other scientific fields, the purely mathematical aspect is met by the engineer's need for sufficiently handling a great number of applications, which are not contained in the typical cases by which the physicist verified this theory. It is, for instance, certainly not enough for geodetic engineering to be able to calculate the "geopotential" for spheres and ellipsoids, but more complicated bodies (e.g., geoidal and/or telluroidal bodies, the real Earth) must be taken into account. The boundary value problem, which provides a mathematical apparatus of these questions, therefore, is not only a purely mathematical subtlety, but also a problem of great practical importance. It is therefore remarkable that the link between mathematics and engineering seems to be much stronger than it is between mathematics and physics. Even more remarkably, engineering (especially, methods of exploration such as gravimetry and magnetometry) encouraged mathematics to deal not only with well-posed problems, but also with ill-posed problems. As a consequence in potential theory, a geodetically relevant example, namely the *inverse gravimetry problem* became attackable (cf. [39, 41]), that aims at the determination of the density contrast function from Newtonian volume potential values on and outside the integration area. Even more, as we already pointed out, the theory of vector spaces and linear operators which originated from the Fredholm theory of boundary value problems created a new mathematical discipline, viz. functional analysis, which provided much of the operator theoretic background of the theory of inverse problems.

Geomathematical Methodology in Inverse Gravimetry In what follows we sketch a novel process for the solution of practical problems of exploration based on gravimetry measurements (for more details the reader is referred to [41]).

3.1.1 "Transfer" from Gravimetric Observation to the Language of Mathematics

Gravity can be measured in different ways. We already know the "Newton principle" of an *absolute gravimeter* (cf. Fig. 3.5).

A *relative gravimeter* compares the value of gravity at one point with another. They measure the ratio of the gravity at the two points. Most common relative gravimeters are spring-based (cf. Fig. 3.6). By determining the amount by which the weight stretches the spring, gravity becomes available (see Figs. 3.6 and 3.7), and the density contrast is reflected by the gravitational values (after reduction of non-gravitative influences).

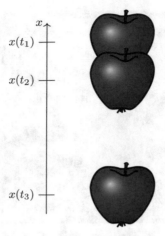

Fig. 3.5 Test mass ("apple") in free fall: the principle of an absolute gravimeter, i.e., the measurements of three time transitions enable the determination of the gravity intensity

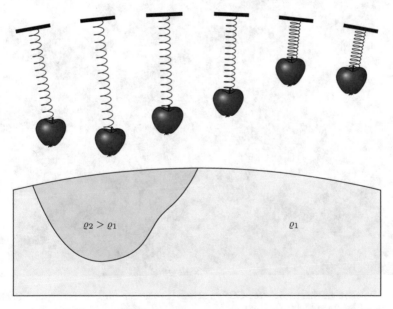

Fig. 3.6 The principle of a relative gravimeter, i.e., the elongation of the spring leading (via Hooke's law) to gravity differences reflects the density distribution inside a test area (from [47])

Fig. 3.7 Relative gravimeter Scintrex-CG6 in action (Landesamt für Vermessung und Geobasis-information Rheinland-Pfalz, Koblenz)

The gravity intensity observed by gravimetry on the Earth's surface (see Figs. 3.7 and 3.8) depends on a number of effects to be removed:

- attraction of the reference surface (e.g., an ellipsoid/spheroid),
- elevation above sea level,
- topography,

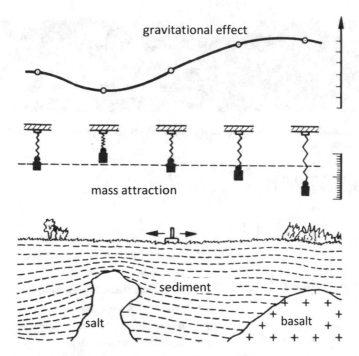

Fig. 3.8 Mass implied gravitational effect obtained by a relative gravimeter (illustration with kind permission of Teubner-publishing taken from [69] in modified form)

- time dependent (tidal) variations,
- (Eötvös) effect of a moving platform,
- isostatic balance on the lower lithosphere,
- density variations inside the upper crust.

In more detail, certain corrections have to be applied to the data in order to account for effects not related to the subsurface: Drift corrections are necessary, since each gravimeter suffers mechanical changes over time, and so does its output measurement. This change is generally assumed to be linear. In case of acquisition on a moving platform, the motion relative to the surface of the Earth implies a change in centrifugal acceleration. The Eötvös correction depends on the latitude and the velocity vector of the moving platform. It should be observed that free air anomaly does not correct for the first two effects which could mask the gravity anomalies related to the Bouguer density contrasts in the crust. Complete Bouguer correction effectively removes the gravity anomalies due to bathymetry, but still contains the gravity effect of the Moho. Isostatics contain the gravity effect of the Moho. Special methods such as Poincare–Bey corrections are in use within boreholes or for special geoid computations.

Fig. 3.9 Illustration of the components of the gravity acceleration (ESA medialab, ESA communication production SP–1314)

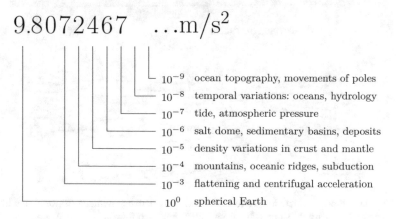

Fig. 3.10 Illustration of the coordinates of the gravity acceleration at a certain location (cf. Fig. 3.9). All gravity measurements are our days expressed in SI units. However, still wide use is made of some older, traditional standards. The unit of gravity acceleration is ms^{-2}, the traditional unit is Gal (after G. Galilei): $1\,\mathrm{Gal} = 0.01\,\mathrm{ms}^{-2}$

As a consequence, to isolate the effects of local density variations (as illustrated in Figs. 3.9 and 3.10 for a special location) from all other contributions, it is necessary to apply a series of reductions:

- The attraction of, e.g., the reference ellipsoid/spheroid has to be subtracted from the measured values.
- An elevation correction must be done, i.e., the vertical gradient of gravity is multiplied by the elevation of the station and the result is added. With increasing elevation of the Earth, there is usually an additional mass between the reference level and the actual level. This additional mass itself exerts a positive gravitational attraction.
- Bouguer correction and terrain correction are applied to correct for the attraction of the slab of material between the observation point and the geoid.
- A terrain correction accounts for the effect of nearby masses above or mass deficiencies below the station. Isostatic correction accounts for the isostatic roots (Moho) (Fig. 3.11).

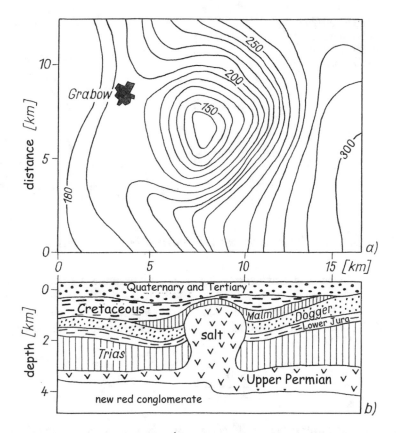

Fig. 3.11 Top: Gravity effect in $[\mu\,m\,s^{-1}]$ of the salt dome Werle (Mecklenburg, Germany); bottom: Geological vertical profile (with kind permission of Teubner-publishing taken from [69] in modified form)

Nowadays, gravimetry is in use all over the world in diverse applications (cf. Figs. 3.8 and 3.12), from which we list only a few examples:

1. Gravimetry is decisive for geodetic purposes of modeling gravity anomalies, geoidal undulations, and quasigeoidal heights.
2. Gravimetry is helpful in different phases of the oil exploration and production processes as well as in geothermal research.
3. Archeological and geotechnical studies aim at the mapping of subsurface voids and overburden variations.
4. Gravimetric campaigns may be applied for groundwater and environmental studies. They help to map aquifers to provide formations and/or structural control.
5. Gravimetric studies give information about tectonically derived changes and volcanological phenomena.

Fig. 3.12 "Taylor mollifiers"
for the Newton kernel in
sectional illustration (shown
are the scales $j = 0, 1, 2, 3$)

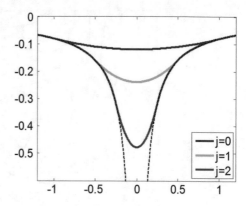

3.1.2 Mathematical Analysis: Direct and Inverse Gravimetry

The aim of the inverse gravimetry problem is to determine the density contrast
function inside a certain subarea $\mathcal{G}$ of the Earth. If the density is supposed to be
a function F of bounded signal energy in $\overline{\mathcal{G}}$, i.e., $\|F\|_{L^2(\mathcal{G})} < \infty$, the gravitational
potential (3.6) (for simplicity, we let $\gamma = 1$ for our theoretical considerations)

$$V(x) = \int_{\mathcal{G}} \underbrace{\frac{1}{4\pi} \frac{1}{|x - y|}}_{=G(\Delta;|x-y|)} F(y)\, dy, \quad x \in \mathbb{R}^3 \tag{3.8}$$

can be calculated everywhere in $\mathbb{R}^3$ according to Newton's famous law (1687), so
that the *direct gravimetry problem*

$$\underbrace{F}_{\substack{= \text{density signature inside } \mathcal{G}}} \longrightarrow \underbrace{V}_{\substack{= \text{gravitational potential in } \mathbb{R}^3}} \tag{3.9}$$

is a matter of (approximate) integration. Already at this early stage, it should be
mentioned that there is a striking dependence in the calculation of a gravitational
potential value $V(x)$ on its position $x \in \mathbb{R}^3$. As far as the point x is situated in the
"outer space" $\mathcal{G}^c = \mathbb{R}^3\backslash\overline{\mathcal{G}}$ ($\overline{\mathcal{G}} = \mathcal{G}\cup\partial\mathcal{G}$, $\partial\mathcal{G}$ boundary surface of $\mathcal{G}$), the value $V(x)$ is
obtained by a proper volume integral. However, for a position x in the "inner space"
$\mathcal{G}$ or on the boundary $\partial\mathcal{G}$, we are confronted with an improper volume integral in
$\mathbb{R}^3$. As a consequence, it may be expected that the *inverse gravimetry problem*

$$\underbrace{V}_{\substack{= \text{gravitational potential in } \mathbb{R}^3\backslash\mathcal{G}}} \longrightarrow \underbrace{F}_{\substack{= \text{density signature in } \mathcal{G}}} \tag{3.10}$$

also shows a striking difference in its solution process and dependence on the position of the gravitational data (cf. [39] for more details). In fact, we recall $-\Delta V(x) = 0$, $x \in \mathbb{R}^3 \backslash \overline{\mathcal{G}}$, whereas $-\Delta V(x) = F(x)$, $x \in \mathcal{G}$.

In accordance with the famous *Hadamard's classification* (cf. [63]), the problem of determining F in $\mathcal{G}$ from potential values $V(x), x \in \mathbb{R}^3 \backslash \overline{\mathcal{G}}$ via the integral equation

$$V(x) = A[F](x) = \int_{\mathcal{G}} F(y) \, G(\Delta; |x - y|) \, dy, \quad F \in L^2(\mathcal{G}) \tag{3.11}$$

violates all criteria of well-posedness, viz. uniqueness, existence, and stability:

1. A solution of the problem exists only if V belongs to the space $Y|\overline{\mathcal{G}^c}$, where $Y = A[L^2(\mathcal{G})]$ is the image of $X = L^2(\mathcal{G})$ under the operator A. However, it should be pointed out that this restriction does not cause any numerical difficulty, since, in practice, the information of V is only finite-dimensional.
2. The most serious problem is the non-uniqueness of the solution: The associated Fredholm integral operator $A_{\overline{\mathcal{G}^c}}$ given by

$$V = A_{\overline{\mathcal{G}^c}}[F], \quad F \in L^2(\mathcal{G}) \tag{3.12}$$

with

$$A_{\overline{\mathcal{G}^c}}[F] := \int_{\mathcal{G}} G(\Delta; |\cdot - y|) \, F(y) \, dy \Big|_{\mathcal{G}^c} \tag{3.13}$$

has a kernel (null space) which is already known to coincide with the $L^2(\mathcal{G})$-orthogonal subspace, which is the closure of all harmonic functions on $\mathcal{G}$. More precisely, if F is a member of class $L^2(\mathcal{G})$, then $A_{\overline{\mathcal{G}^c}} : L^2(\mathcal{G}) \to Y|\mathcal{G}^c$ defines a linear operator such that $A_{\overline{\mathcal{G}^c}}[F]$ is continuous in $\overline{\mathcal{G}^c}$, harmonic in $\mathcal{G}^c$, and regular at infinity. The operator $A_{\overline{\mathcal{G}^c}}$ as defined by (3.12) is surjective, but it is not injective. Indeed, the null space (kernel) of $A_{\overline{\mathcal{G}^c}}$, i.e.,

$$\mathcal{N}(A_{\overline{\mathcal{G}^c}}) = \text{AnHarm}(\mathcal{G}) \tag{3.14}$$

consists of all functions in $L^2(\mathcal{G})$ that are orthogonal to the space $\text{Harm}(\mathcal{G})$ of harmonic functions in $\mathcal{G}$. $\mathcal{N}(A_{\overline{\mathcal{G}^c}})$ is the space of anharmonic functions in $\mathcal{G}$. In fact, we have (see, e.g., [82, 121])

$$L^2(\mathcal{G}) = \overline{\text{Harm}(\mathcal{G})}^{\|\cdot\|_{L^2(\mathcal{G})}} \oplus \left(\overline{\text{Harm}(\mathcal{G})}^{\|\cdot\|_{L^2(\mathcal{G})}} \right)^{\perp}, \tag{3.15}$$

hence,

$$L^2(\mathcal{G}) = \text{Harm}(\mathcal{G}) \oplus \text{AnHarm}(\mathcal{G}) = \text{Harm}(\mathcal{G}) \oplus \mathcal{N}(A_{\overline{\mathcal{G}^c}}). \tag{3.16}$$

Unfortunately, the orthogonal complement, i.e., the class of *anharmonic functions*, is infinite-dimensional (see, e.g., [82, 84, 121]).

3. Restricting the operator $A_{\overline{\mathcal{G}^c}}$ to $\mathrm{Harm}(\mathcal{G})$ leads to an injective mapping which, however, has a discontinuous inverse.

Concerning the historical background, the problem of non-uniqueness has been discussed extensively in the literature (for a more detailed analysis see, e.g., [84]). This problem can be resolved by imposing some reasonable additional condition on the density. As we already saw, a suitable condition, suggested by the mathematical structure of the Newton potential operator A, is to require that the density be harmonic. In fact, the approximate calculation of a harmonic density has already been implemented in several papers (see, e.g., [86, 88] and the references therein), whereas the problem of determining the anharmonic part seems to be still a great challenge.

Altogether, it should be remarked that up to now *the ill-posedness of the inverse gravimetry problem seriously limits its application in geoscience and exploration*, but on and off in geodesy (see, e.g., [117]) and particularly in geothermal research (see, e.g., [43]), we are able to take advantage of gravitational data systems inside the area $\mathcal{G}$ under the specific consideration. However, under the knowledge of additional internal gravimeter data, the methodological situation will change drastically, and significant improvement may be expected for practical applicability.

3.1.3 Development of a Mathematical Solution Method

Intuitively, given a function which is rather irregular, by convolving it with a *mollifier* the function gets "mollified," that is, its sharp features are smoothed. The mollifiers used here in inverse gravimetry (cf. [39]) are smooth functions with special properties to create sequences of smooth functions approximating the Newton kernel. Our development actually aims at a twofold mollification (cf. [39, 41]) for which the convolutions, i.e., the mollified potential and density approximations, are related in the sense of Poisson's differential equation by the negative Laplace derivative. To be more precise, within the potential framework, the Newton kernel is mollified by certain Taylor expansions so that the negative Laplace derivative of the Taylor expressions within the contrast signature framework leads to mollifiers of the Dirac kernel, i.e., singular integrals. In doing so, a substitute of the potential simultaneously provides an associated approximation of the density (for numerical tests and experiments see [10, 85]).

Multiscale Mollifier Technique More concretely, by one-dimensional Taylor linearization we obtain

$$\frac{1}{\sqrt{u}} = \frac{1}{\sqrt{u_0}} - \frac{1}{2}\frac{1}{u_0^{\frac{3}{2}}}(u - u_0) + \frac{3}{8}\frac{1}{(u_0 + \vartheta(u - u_0))^{\frac{5}{2}}}(u - u_0)^2 \qquad (3.17)$$

for some $\vartheta \in (0, 1)$. Setting $u = r^2$ and $u_0 = \varrho^2$ we therefore find

$$\frac{1}{r} = \frac{1}{2\varrho} \left(3 - \frac{r^2}{\varrho^2} \right) + \frac{3}{8} \frac{1}{(\varrho^2 + \vartheta (r^2 - \varrho^2))^{\frac{5}{2}}} (r^2 - \varrho^2)^2. \tag{3.18}$$

In other words, by letting $r = |x - y|$ we are able to give a simple example for a "mollification" of the fundamental solution of the Laplace equation

$$G(\Delta; r) := \frac{1}{4\pi r}, \qquad r > 0, \tag{3.19}$$

by

$$G_\varrho^H (\Delta; r) := \begin{cases} \dfrac{1}{8\pi\varrho} \left(3 - \dfrac{1}{\varrho^2} r^2 \right), & r \leq \varrho \\[4mm] \dfrac{1}{4\pi r}, & r > \varrho. \end{cases} \tag{3.20}$$

such that $G_\varrho^H (\Delta; \cdot)$ is continuously differentiable for all $r \geq 0$. Obviously, $G(\Delta; r) = G_\varrho^H (\Delta; r)$ for all $r > \varrho$. As a consequence,

$$G(\Delta; |x - y|) = \frac{1}{4\pi} \frac{1}{|x - y|}, \qquad |x - y| \neq 0, \tag{3.21}$$

admits a *"mollification" (regularization)* of the form

$$G_\varrho^H (\Delta; |x - y|) = \begin{cases} \dfrac{1}{8\pi\varrho} \left(3 - \dfrac{1}{\varrho^2} |x - y|^2 \right), & |x - y| \leq \varrho \\[4mm] \dfrac{1}{4\pi} \dfrac{1}{|x - y|}, & \varrho < |x - y|. \end{cases} \tag{3.22}$$

Let $F : \overline{\mathcal{G}} \to \mathbb{R}$ be of class $C^{(0)} (\overline{\mathcal{G}})$. We set

$$V_\varrho^H (x) = \int_{\mathcal{G}} G_\varrho^H (\Delta; |x - y|) F(y) \, dy. \tag{3.23}$$

The integrands of V and V^ϱ differ only in the ball $\mathbb{B}_\varrho (x)$ around the point x with radius ϱ, i.e., $\mathbb{B}_\varrho (x) = \{ y \in \mathbb{R}^3 : |x - y| < \varrho \}$. Because of its continuity, the function $F : \overline{\mathcal{G}} \to \mathbb{R}$ is uniformly bounded on $\overline{\mathcal{G}}$. This fact shows that

$$\left| V(x) - V_\varrho^H (x) \right| = O \left(\int_{\mathbb{B}_\varrho (x)} |G(\Delta; |x - y|) - G_\varrho^H (\Delta; |x - y|)| \, dy \right) = O(\varrho^2), \tag{3.24}$$

$\varrho \to 0$ (cf. [46]). Therefore we additionally obtain that V is of class $C^{(0)}(\overline{\mathcal{G}})$ as the limit of a uniformly convergent sequence of continuous functions on $\overline{\mathcal{G}}$.

We continue with the differential equation

$$- \Delta_y \, G_\varrho^H(\Delta; |y - z|) = \; H_\varrho(|y - z|) \tag{3.25}$$

with

$$G_\varrho^H(\Delta; |y - z|) = \begin{cases} \frac{1}{8\pi\varrho}(3 - \frac{|y-z|^2}{\varrho^2}) \, , \, |y - z| \leq \varrho, \\ \frac{1}{4\pi|y-z|} \qquad\quad , \, |y - z| > \varrho, \end{cases} \tag{3.26}$$

where

$$H_\varrho(|y - z|) = \begin{cases} \frac{3}{4\pi\varrho^3} \, , \, |y - z| \leq \varrho, \\ 0 \qquad , \, |y - z| > \varrho \end{cases} \tag{3.27}$$

is the so-called *Haar kernel* (note that $\|\mathbb{B}_\varrho(0)\| = \frac{4}{3}\pi\varrho^3$). It is well known (see, e.g., [30]) that the *Haar singular integral* $\{I_\varrho\}_{\varrho>0}$ defined by

$$F_\varrho^H := I_\varrho[F] = \int_{\mathcal{G}} H_\varrho(|\cdot -z|) \, F(z) \, dz, \tag{3.28}$$

with the *Haar kernel as mollifier* satisfies the limit relation

$$\lim_{\varrho \to 0+} I_\varrho[F] = F, \quad F \in L^2(\mathcal{G}), \tag{3.29}$$

in the topology of $L^2(\mathcal{G})$. Moreover, we have

$$\lim_{\varrho \to 0+} I_\varrho[F](x) = \alpha(x) \, F(x), \quad x \in \overline{\mathcal{G}}, \; F \in C^{(0)}(\overline{\mathcal{G}}), \tag{3.30}$$

where $\alpha(x)$ is the solid angle at x subtended by the surface $\partial\mathcal{G}$.

Correspondingly to $\{I_\varrho\}_{\varrho>0}$ we introduce the family $\{A_\varrho\}_{\varrho>0}$ given by

$$V_\varrho^H := A_\varrho^H[F] = \int_{\mathcal{G}} G_\varrho^H(\Delta; |\cdot -z|) \, F(z) \, dz, \tag{3.31}$$

such that

$$\Delta \, A_\varrho^H[F] = \Delta \int_{\mathcal{G}} G_\varrho^H(\Delta; |\cdot -z|) \, F(z) \, dz \tag{3.32}$$

$$= -I_\varrho[F] = -F_\varrho^H$$

$$= -\int_{\mathcal{G}} H_\varrho(|\cdot -z|) \, F(z) \, dz.$$

Of course, the Haar kernel may be replaced by any other singular integral kernel (cf. [39]). For simplicity, however, we confine our considerations to the Haar kernel, although its discontinuity is not of advantage for most of the numerical computations.

We are interested in applying the multiscale "Haar philosophy" to an approximate determination of the mass density distribution inside $\mathcal{G}$ (see [39] for the conceptual background).

Suppose that $\{\varrho_j\}_{j \in \mathbb{N}_0}$ is a positive, monotonously decreasing sequence with $\lim_{j \to \infty} \varrho_j = 0$, for example, the dyadic sequence given by $\varrho_j = 2^{-j}$. For $j \in \mathbb{N}_0$, we consider the differences

$$\Psi_{G^H_{\varrho_j}}(\Delta; |\cdot - y|) = G^H_{\varrho_{j+1}}(\Delta; |\cdot - y|) - G^H_{\varrho_j}(\Delta; |\cdot - y|) \tag{3.33}$$

and

$$\Psi_{H_{\varrho_j}}(|\cdot - y|) = H_{\varrho_{j+1}}(|\cdot - y|) - H_{\varrho_j}(|\cdot - y|). \tag{3.34}$$

$\Psi_{G^H_{\varrho_j}}(\Delta; \cdot)$ and $\Psi_{H^{\varrho_j}}$ are called "ϱ_j-*fundamental wavelet function*" and "ϱ_j-*Haar wavelet function*," respectively.

The associated "ϱ_j-*potential wavelet functions*" and the "ϱ_j-*density wavelet functions*" are given by

$$(WV)^H_{\varrho_j} = \int_{\mathcal{G}} \Psi_{G^H_{\varrho_j}}(\Delta; |\cdot - y|) \, F(y) \, dy \tag{3.35}$$

and

$$(WF)^H_{\varrho_j} = \int_{\mathcal{G}} \Psi_{H_{\varrho_j}}(|\cdot - y|) \, F(y) \, dy, \tag{3.36}$$

respectively. The ϱ_j-potential wavelet functions and the ϱ_j-density wavelet functions, respectively, characterize the successive *detail information* contained in $V^H_{\varrho_{j+1}} - V^H_{\varrho_j}$ and $F^H_{\varrho_{j+1}} - F^H_{\varrho_j}$, $j \in \mathbb{N}_0$. In other words, we are able to decorrelate the potential V and the "density signature" F, respectively, in the form of "band structures"

$$(WV)^H_{\varrho_j} = V^H_{\varrho_{j+1}} - V^H_{\varrho_j}, \tag{3.37}$$

and

$$(WF)^H_{\varrho_j} = F^H_{\varrho_{j+1}} - F^H_{\varrho_j}. \tag{3.38}$$

The essential problem to be solved in multiscale extraction of geological features is to identify those detail information, i.e., band structures in (3.37), which contain specifically desired geological (density) characteristics in (3.38). Seen from a numerical point of view, it is remarkable that both wavelet functions $y \mapsto \Psi_{G^H_{\varrho_j}}(\Delta; |\cdot - y|)$ and $y \mapsto \Psi_{H_{\varrho_j}}(|\cdot - y|)$ vanish outside a ball around the center x due to their construction, i.e., these functions are spacelimited showing a ball as local support. Furthermore, the ball becomes smaller with increasing scale parameter j, so that more and more high-frequency phenomena can be highlighted without changing the features outside the balls.

Forming the telescoping sums

$$\sum_{j=0}^{J-1} (WV)^H_{\varrho_j} = \sum_{j=0}^{J-1} \left(V^H_{\varrho_{j+1}} - V^H_{\varrho_j} \right), \tag{3.39}$$

and

$$\sum_{j=0}^{J-1} (WF)^H_{\varrho_j} = \sum_{j=0}^{J-1} \left(F^H_{\varrho_{j+1}} - F^H_{\varrho_j} \right), \tag{3.40}$$

we easily arrive at the scale-discrete identities

$$V^H_{\varrho_J} = V^H_{\varrho_0} + \sum_{j=0}^{J-1} (WV)^H_{\varrho_j} \tag{3.41}$$

and

$$F^H_{\varrho_J}(x) = F^H_{\varrho_0} + \sum_{j=0}^{J-1} (WF)^H_{\varrho_j}. \tag{3.42}$$

Thus we finally end up with the following multiscale relations:

$$\lim_{J \to \infty} V^H_{\varrho_J} = V^H_{\varrho_0} + \sum_{j=0}^{\infty} (WV)^H_{\varrho_j} \tag{3.43}$$

and

$$\lim_{J \to \infty} F^H_{\varrho_J} = F^H_{\varrho_0} + \sum_{j=0}^{\infty} (WF)^H_{\varrho_j} = - \lim_{J \to \infty} \Delta V^H_{\varrho_J} = -\Delta V^H_{\varrho_0} - \sum_{j=0}^{\infty} \Delta(WV)^H_{\varrho_j}. \tag{3.44}$$

Altogether, the potential V as well as the "density signature" F can be expressed by initial low-pass filtered signals $V^H_{\varrho_0}$ and $F^H_{\varrho_0}$ and successive addition of band-pass filtered signals $(WV)^H_{\varrho_j}$ and $(WF)^H_{\varrho_j}$, $j = 0, 1, \dots,$ respectively.

It should be mentioned that our multiscale approach is constructed such that, within the spectrum of all wavebands (see (3.37) and (3.38)), certain rock formations or aquifers, respectively, may be associated with a specific band characterizing typical features within the multiscale reconstruction (cf. [10]). Each scale parameter in the decorrelation is assigned to correspond to a low-pass approximation of the data at a particular resolution. The wavelet contributions are obtained as part within a multiscale approximation by calculating the difference between two consecutive scaling functions. In other words, the wavelet transformation (filtering) of a signal constitutes the difference of two low-pass filters, thus it may be regarded as a band-pass filter. Due to our construction, the wavelets show an increasing space localization as the scale increases. In this way, the characteristic signatures of a signal can be detected in certain frequency bands.

For a sufficiently large integer J and $F \in C^{(0)}(\overline{\mathcal{G}})$ it follows from (3.30) that

$$\alpha(x)\, F(x) \cong I_{\varrho_J}[F](x) = F_{\varrho_J}^H(x) = \int_{\mathcal{G}} H_{\varrho_J}(|x - y|) F(y)\, dy, \quad x \in \overline{\mathcal{G}}, \qquad (3.45)$$

("$\cong$"means that the error is negligible). From (3.25) we obtain

$$\Delta\, A_{\varrho_J}^H[F](x) = \Delta_x \int_{\mathcal{G}} G_{\varrho_J}^H(\Delta; |x - y|) F(y)\, dy \qquad (3.46)$$

$$= - \int_{\mathcal{G}} H_{\varrho_J}(|x - y|) F(y)\, dy$$

$$\cong - \alpha(x)\, F(x),$$

where we are aware of the fact that

$$V(x) \cong \int_{\mathcal{G}} G_{\varrho_J}^H(\Delta; |x - y|) F(y)\, dy, \quad x \in \mathbb{R}^3, \qquad (3.47)$$

with negligible error.

In order to realize a fully discrete approximation of F we have to apply approximate integration formulas leading to

$$V(x) \cong \sum_{i=1}^{N_J} G_{\varrho_J}^H(\Delta; |x - y_i^{N_J}|)\, w_i^{N_J} F(y_i^{N_J}), \quad x \in \mathbb{R}^3, \qquad (3.48)$$

where $w_i^{N_J} \in \mathbb{R}$, $y_i^{N_J} \in \overline{\mathcal{G}}$, $i = 1, \ldots, N_J$, are the known weights and knots, respectively.

For numerical realization of mass density modeling by means of Haar kernels we notice that all coefficients $a_i^{N_J} = w_i^{N_J} F(y_i^{N_J})$, $i = 1, \ldots, N_J$, are unknown. Then we have to solve a linear system, namely

$$V(x_k^{T_J}) = \sum_{i=1}^{N_J} G_{\varrho_J}^H(\Delta; |x_k^{T_J} - y_i^{N_J}|) \, a_i^{N_J}, \quad x_k^{T_J} \in \overline{\mathcal{G}}, \, k = 1, \ldots, T_J, \qquad (3.49)$$

to determine $a_i^{N_J}$, $i = 1, \ldots, N_J$, from known gravitational values $V(x_k^{T_J})$ at knots $x_k^{T_J} \in \mathbb{R}^3$, $k = 1, \ldots, T_J$. Once all density values $F(y_i^{N_J})$, $i = 1, \ldots, N_J$, are available (note that the integration weights $w_i^{N_J}$, $i = 1, \ldots, N_J$, are known), the density distribution F can be obtained from the formula

$$F(x) \cong F_{\varrho_J}^H(x) = \sum_{i=1}^{N_J} H_{\varrho_J}(|x - y_i^{N_J}|) \, \underbrace{w_i^{N_J} F(y_i^{N_J})}_{=a_i^{N_J}}, \quad x \in \mathbb{R}^3. \qquad (3.50)$$

Even better, fully discrete Haar filtered versions of F at lower scales can be derived in accordance with the approximate integration rules

$$\int_{\mathcal{G}} H_{\varrho_j}(|x-z|) \, F(z) \, dV(z) \cong \sum_{i=1}^{N_j} H_{\varrho_j}(|x-y_i^{N_j}|) \, w_i^{N_j} \, F(y_i^{N_j}), \quad x \in \mathcal{G}, \qquad (3.51)$$

for $j = J_0, \ldots, J$, where $w_i^{N_j}, y_i^{N_j}, i = 1, \ldots, N_j$, are known weights and knots, respectively, such that $\{y_1^{N_j}, \ldots, y_{N_j}^{N_j}\} \subset \{y_1^{N_J}, \ldots, y_{N_J}^{N_J}\} \subset \overline{\mathcal{G}}$, i.e., the sequence of knots $\{y_1^{N_J}, \ldots, y_{N_J}^{N_J}\} \subset \overline{\mathcal{G}}$ shows a hierarchical positioning.

Altogether, our approach yields Haar filtered versions establishing a (space-based) multiscale decomposition $F_{\varrho_J}^H, \ldots, F_{\varrho_{J_0}}^H$ of the density distribution F, such that an entire set of approximations is available from a single locally supported "mother function," i.e., the Haar kernel function (3.27), and this set provides useful "building block functions," which enable decorrelation of the density signatures and suitable storage and fast decorrelation of density data. Moreover, fully discrete Haar filtered versions of F at lower scales can be derived in accordance with the approximate integration rules

$$F_{\varrho_j}^H(x) = \int_{\mathcal{G}} H_{\varrho_j}(|x - y|) \, F(y) \, dy \cong \sum_{i=1}^{N_j} H_{\varrho_j}(|x - y_i^{N_j}|) \, w_i^{N_j} \, F(y_i^{N_j}), \quad x \in \mathcal{G},$$

$$(3.52)$$

for $j = J_0, \ldots, J$, where $w_i^{N_j}, y_i^{N_j}, i = 1, \ldots, N_j$, are known weights and knots, respectively, such that we can take advantage of the fact that $\{y_1^{N_j}, \ldots, y_{N_j}^{N_j}\} \subset \{y_1^{N_J}, \ldots, y_{N_J}^{N_J}\} \subset \overline{\mathcal{G}}$.

The serious problem of our multiscale approach is that measurements of gravitation are only available in the interior $\mathcal{G}$ in exceptional cases, for example, locally in geothermal boreholes. Usually, we are able to take into account surface measurements on $\partial \mathcal{G}$, but it may be questioned in view of the ill-posedness that deep geological formations can be detected by an exclusive use of terrestrial gravitational data. As a consequence, we did not use exclusively terrestrial data in our numerics. Instead, we designed an iterative procedure for adding subsequently from step to step more and more discrete internal potential values. In doing so, for a test area in the Bavarian Molasse area, we obtained good results which could be validated by geological information known from reflection seismics.

Mollifier Inversion Methods We present some methods for the regularization of the inverse gravimetry problem by mollifiers (for more detail see [39]).

Method 1: Mollified Numerical Integration For the numerical realization of mass density modeling by means of Haar kernels we come back to the linear system (see (3.49))

$$
V(x_k^{T_J}) \cong \sum_{i=1}^{N_J} G_{\varrho J}^H(\Delta; |x_k^{T_J} - y_i^{N_J}|)\, a_i^{N_J}, \quad x_k^{T_J} \in \overline{\mathcal{G}}, \ k = 1, \ldots, T_J, \qquad (3.53)
$$

to determine $a_i^{N_J}, i = 1, \ldots, N_J$, from known gravitational values $V(x_k^{T_J})$ at knots $x_k^{T_J} \in \mathbb{R}^3, k = 1, \ldots, T_J$. Obviously,

$$
V_{\varrho j}^H(x) \cong \sum_{i=1}^{N_J} G_{\varrho j}^H(\Delta; |x - y_i^{N_J}|) \underbrace{w_i^{N_J} F(y_i^{N_J})}_{=a_i^{N_J}}, \quad x \in \mathcal{G}, \qquad (3.54)
$$

with

$$
G_{\varrho j}^H(\Delta; |x - y|) = \begin{cases} \dfrac{1}{8\pi \varrho_j}\left(3 - \dfrac{1}{\varrho_j{}^2}|x - y|^2\right), & |x - y| \le \varrho_j \\[4mm] \dfrac{1}{4\pi}\dfrac{1}{|x - y|}, & \varrho_j < |x - y|. \end{cases} \qquad (3.55)
$$

Once all density values $F(y_i^{N_J})$, $i = 1, \ldots, N_J$, are available (note that the integration weights $w_i^{N_J}$, $i = 1, \ldots, N_J$, are known), the density distribution F can be obtained from the formula

$$F(x) \cong F_{\varrho_J}^H(x) = \sum_{i=1}^{N_J} H_{\varrho_J}(|x - y_i^{N_J}|) \underbrace{w_i^{N_J} F(y_i^{N_J})}_{=a_i^{N_J}}, \quad x \in \mathbb{R}^3. \tag{3.56}$$

Even better, fully discrete Haar filtered versions of F in $\mathcal{G}$ at lower scales can be derived in accordance with the rules

$$F_{\varrho_j}^H(x) = \int_{\mathcal{G}} H_{\varrho_j}(|x - z|) \, F(z) \, dV(z) \cong \sum_{i=1}^{N_J} H_{\varrho_j}(|x - y_i^{N_J}|) \underbrace{w_i^{N_J} F(y_i^{N_J})}_{=a_i^{N_J}}, \quad x \in \mathcal{G},$$

$$\tag{3.57}$$

for $j = J_0, \ldots, J$.

Altogether, our approach yields Haar filtered versions establishing a (space-based) multiscale decomposition $F_{\varrho_J}^H, \ldots, F_{\varrho_{J_0}}^H$ of the density distribution F, such that an entire set of approximations is available from a single locally supported "mother function," i.e., the Haar kernel function (3.27), and this set provides useful "building block functions," which enable decorrelation of the density signatures

$$(WF)_{\varrho_j}^H(x) = \int_{\mathcal{G}} \Psi_{H_{\varrho_j}}(|x - y|) \, F(y) \, dy \cong \sum_{i=1}^{N_J} \Psi_{H_{\varrho_j}}(|x - y_i^{N_J}|) \underbrace{w_i^{N_J} F(y_i^{N_J})}_{=a_i^{N_J}}.$$

$$\tag{3.58}$$

Method 2: Spline Interpolation/Smoothing We denote by Y the space of all *Newton integrals* $A[F]$ (cf. [39]) given by

$$A[F] = \int_{\mathcal{G}} G(\Delta; |x - y|) \, F(y) \, dy, \quad F \in L^2(\mathcal{G}), \tag{3.59}$$

with $G(\Delta; |\cdot - \cdot|)$ given by (3.72). In other words,

$$\mathrm{Y} = A[L^2(\mathcal{G})]. \tag{3.60}$$

The image space $\mathrm{Y} = A[L^2(\mathcal{G})]$ *is a reproducing kernel Hilbert space possessing the reproducing kernel* (see [3, 16, 66, 109] for more details on reproducing Hilbert space structure)

$$K_{\mathrm{Y}}(x, y) = \int_{\mathcal{G}} G(\Delta; |x - z|) \, G(\Delta; |z - y|) \, dz, \quad x, y \in \mathbb{R}^3 \tag{3.61}$$

The reproducing property guarantees that the convolution of the kernel (3.61) with a potential $P \in Y$ (in the Y-topology) "reproduces" to P, so that no numerical integration procedure is needed for the operation of the convolution. Nevertheless, it remains to calculate the integral in (3.61) by approximate integration, i.e., cubature rules.

Finally, it should be mentioned that

$$- \Delta_x K_Y(x, y) = G(\Delta; |x - y|) \, dz, \quad x \neq y, \quad x, y \in \mathbb{R}^3. \tag{3.62}$$

In Y minimum norm (spline) interpolation can be performed in the usual way (see, e.g., [40]):

Let $\{x_1, \ldots, x_N\} \subset \mathbb{R}^3$ be a point system with $x_i \neq x_k, i \neq k$. Then, within the set

$$\mathcal{I}_{x_1,\ldots,x_N}^V = \{U \in Y : U(x_i) = V(x_i) = \gamma_i, \, i = 1, \ldots, N\}, \tag{3.63}$$

the minimum norm interpolation problem of finding S_N^V that satisfies

$$\|S_N^V\|_Y = \inf_{U \in \mathcal{I}_{x_1,\ldots,x_N}^V} \|U\|_Y \tag{3.64}$$

is well posed, i.e., its solution exists, is unique, and depends continuously on the data $\gamma_1, \ldots, \gamma_N$.

The uniquely determined solution S_N^V is given in the explicit form

$$S_N^V(x) = \sum_{i=1}^N a_i^N K_Y(x, x_i), \quad x \in \mathbb{R}^3, \tag{3.65}$$

where the coefficients $a_1^N, \ldots, a_N^N$ have to be determined by solving the linear system of equations

$$\sum_{i=1}^N a_i^N K_Y(x_k, x_i)] = \gamma_k, \quad k = 1, \ldots, N. \tag{3.66}$$

(note that the matrix $(K_Y(x_k, x_k))_{i,k=1,\ldots,N}$ is as Gram matrix of linearly independent functions $K_Y(x_k, \cdot), k = 1, \ldots, N$, positive definite).

As a result we obtain an approximation of the density distribution as a linear combination of Newton kernels, which are harmonic for $x \in \mathbb{R}^3 \setminus \{x_1, \ldots, x_N\}$:

$$\varrho_N^V(x) = -\Delta S_n^V(x) = -\sum_{i=1}^N a_i^N G(\Delta; |x - x_i|), \, i = 1, \ldots, N. \tag{3.67}$$

In fact, the critical point in the aforementioned spline approach is the solution of the linear system (3.66). Each coefficient $K_Y(x_k, x_i)$ must be determined by numerical integration and the coefficient matrix $\{K_Y(x_k, x_i)\}$ is full-sized. Moreover, numerical instability comes from the increasing "correlation" between the "constituting kernel functions." This can be easily seen by computing the cosine of the angle between two of them, as a correlation coefficient (cf. [45])

$$\varrho_{\text{corr}} = \frac{\langle K_Y(x_k, \cdot), \ K_Y(x_i, \cdot)\rangle_Y}{\| K_Y(x_k, \cdot) \|_Y \| K_Y(x_i, \cdot) \|_Y} = \frac{K_Y(x_k, x_i)}{\sqrt{K_Y(x_k, x_k)K_Y(x_i, x_i)}}. \tag{3.68}$$

It follows that $\varrho_{\text{corr}} \to 1$ when x_k, x_i become closer and closer. To implement bases that reduce, or even annihilate, such correlation is exactly the idea underlying the construction of a wavelet scheme. These numerical calamities led Freeden and Nashed [41] either to decorrelate the reproducing kernel $K_Y(\cdot, \cdot)$ by Gaussian bell functions or to replace $K_Y(\cdot, \cdot)$ by mollifier spline-wavelets. The forthcoming study can be regarded as basic background material for mollifier spline-wavelets, however, only for the special case of spacelimited Haar mollifiers (for more general mollifiers see [41]).

Gaussian Sum Representation Our interest is in substituting the kernel of a "monopole" by a linear combination of Gaussians, i.e.,

$$\frac{1}{|z - x|} \cong \sum_{m=1}^{M} \omega_m e^{-\alpha_m |x-z|^2}. \tag{3.69}$$

The critical point concerning (3.69) is to specify the coefficients $\alpha_m, \omega_m, \ m = 1, \dots, M$, and the integer M.

Different approaches have been proposed in the literature:

- In [61], the approximation is attacked by a Newton-type optimization procedure.
- In [62], a Remez algorithm exploits specific properties of a certain error functional (not discussed here in more detail).
- Fast multipole methods (see, e.g., [12, 56, 58–60]) also provide tools of great numerical significance.

Our approach described in [39] closely parallels the concepts presented in [8, 9]. This concept numerically realized in [11] starts with an initial approximation obtained by the appropriate discretization of an integral expression of $|z - x|^{-1}$. Afterwards, in order to reduce the number M of terms of the Gaussian sum on the right side of (3.69), an algorithm is applied based on Prony's method.

An advantage is that one is able to develop our results for the one-dimensional function $r \mapsto r^{-1}$, $r \in [\delta, 1]$, with some sufficiently small $\delta > 0$:

$$\frac{1}{r} \cong \sum_{m=1}^{M} \omega_m e^{-\alpha_m r^2}, \quad r \in [\delta, 1], \tag{3.70}$$

so that

$$\int_{\mathcal{G}} \frac{1}{|x - z||y - z|} \, dz \cong \sum_{m=1}^{M} \sum_{n=1}^{M} \omega_m \omega_n \int_{\mathcal{G}} e^{-\alpha_m |x-z|^2} e^{-\alpha_n |y-z|^2} \, dz \qquad (3.71)$$

hence, the right-hand side of (3.71) allows the separation into Cartesian coordinates; thus, multi-dimensional integrals over cuboids can be handled iteratively, just by one-dimensional integration techniques. In the case of a ball $\mathcal{G}$, the right side of the integral (3.71) can be calculated by elementary manipulations (cf. [11]).

Furthermore, we are allowed to express $G_{\varrho_j}^H(\Delta; |x - y|)$ in the following way:

$$G_{\varrho_j}^H(\Delta; |x - y|) \cong \begin{cases} \dfrac{1}{8\pi \varrho_j} \left(3 - \dfrac{1}{\varrho_j^{\,2}} |x - y|^2 \right), & |x - y| \le \varrho_j \\[2em] \dfrac{1}{4\pi} \displaystyle\sum_{m=1}^{M} \omega_m e^{-\alpha_m |x-y|^2}, & \varrho_j < |x - y|. \end{cases} \qquad (3.72)$$

Method 3: Mollified Spline Interpolation/Smoothing We denote by Y_{ϱ_J} the space of all *mollified Newton integrals* $A_{\varrho_J}^H[F]$ given by

$$A_{\varrho_J}^H[F] = \int_{\mathcal{G}} G_{\varrho_J}^H(\Delta; |x - y|) \, F(y) \, dy, \quad F \in L^2(\mathcal{G}), \qquad (3.73)$$

with $G_{\varrho_J}^H(\Delta; | \cdot - \cdot |)$ given by (3.72). In other words,

$$Y_{\varrho_J} = A_{\varrho_J}^H[L^2(\mathcal{G})]. \qquad (3.74)$$

From [39] we borrow the following results:
The image space $Y_{\varrho_J} = A_{\varrho_J}^H[L^2(\mathcal{G})]$ is a reproducing kernel Hilbert space possessing the reproducing kernel

$$K_{Y_{\varrho_J}}(x, y) = \int_{\mathcal{G}} G_{\varrho_J}^H(\Delta; |x - z|) \, G_{\varrho_J}^H(\Delta; |z - y|) \, dz, \quad x, y \in \mathbb{R}^3 \qquad (3.75)$$

The reproducing property guarantees that the convolution of the kernel (3.75) with a potential $P \in Y_{\varrho_J}$ (in the Y_{ϱ_J}-topology) "reproduces" to P, so that no numerical integration procedure is needed for the operation of the convolution. Nevertheless, it remains to calculate the integrals in (3.71) by approximate integration methods.

Finally, it should be mentioned that

$$- \Delta_x \, K_{\mathrm{Y}_{\varrho_J}}(x, y) = \int_{\mathcal{G}} H_{\varrho_J}(|x - z|) \, G^H_{\varrho_J}(\Delta; |z - y|) \; dz, \quad x, y \in \mathbb{R}^3. \qquad (3.76)$$

In Y_{ϱ_J}, minimum norm (spline) interpolation can be performed in the usual way (see, e.g., [40]):

Let $\{x_1, \ldots, x_N\} \subset \mathbb{R}^3$ be a point system with $x_i \neq x_k, i \neq k$. Then, within the set

$$\mathcal{I}^V_{x_1,\ldots,x_N} = \{U \in \mathrm{Y}_{\varrho_J} : U(x_i) = V(x_i) = \gamma_i, \; i = 1, \ldots, N\}, \qquad (3.77)$$

the minimum norm interpolation problem of finding S^V_N that satisfies

$$\|S^V_N\|_{\mathrm{Y}_{\varrho_J}} = \inf_{U \in \mathcal{I}^V_{x_1,\ldots,x_N}} \|U\|_{\mathrm{Y}_{\varrho_J}} \qquad (3.78)$$

is well posed, i.e., its solution exists, is unique, and depends continuously on the data $\gamma_1, \ldots, \gamma_N$.

The uniquely determined solution S^V_N is given in the explicit form

$$S^V_N(x) = \sum_{i=1}^N a^N_i K_{\mathrm{Y}_{\varrho_J}}(x, x_i), \quad x \in \mathbb{R}^3, \qquad (3.79)$$

where the coefficients $a^N_1, \ldots, a^N_N$ have to be determined by solving the linear system of equations (for combined interpolation/smoothing see [48])

$$\sum_{i=1}^N a^N_i K_{\mathrm{Y}_{\varrho_J}}(x_k, x_i)] = \gamma_k, \quad k = 1, \ldots, N. \qquad (3.80)$$

(note that the matrix $(K_{\mathrm{Y}_{\varrho_J}}(x_k, x_i))_{i,k=1,\ldots,N}$ is as Gram matrix of linearly independent functions $K_{\mathrm{Y}_{\varrho_J}}(x_i, \cdot), k = 1, \ldots, N$, positive definite).

Now we obtain an approximation of the density distribution as a linear combination of integral-type kernels (3.76) which are not harmonic:

$$\varrho^V_N(x) = -\Delta_x S^V_N(x) = -\sum_{i=1}^N a^N_i \Delta_x K_{\mathrm{Y}_{\varrho_J}}(x, x_i,), \; i = 1, \ldots, N. \qquad (3.81)$$

For purposes of decorrelation, we are indeed able to take advantage of the sparse character of the resulting wavelets to reduce considerably the computational effort (Fig. 3.13).

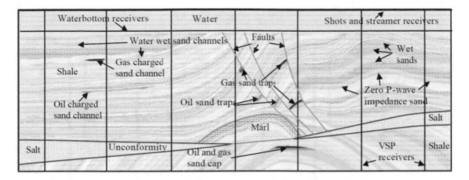

Waterbottom receivers	Water		Shots and streamer receivers
Water wet sand channels	Faults		Wet sands
Shale	Gas charged sand channel		
		Gas sand traps	Zero P-wave impedance sand
Oil charged sand channel	Oil sand traps		Salt
		Marl	
Salt	Unconformity	Oil and gas sand cap	VSP receivers	Shale

Fig. 3.13 Marmousi density model and its geological interpretation (following [78, 116])

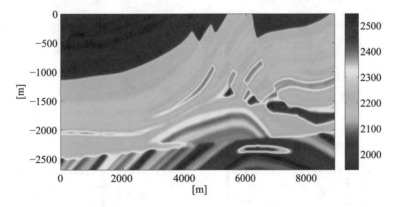

Fig. 3.14 Cross-section of the 3D-Marmousi density model (following [78, 116])

3.1.4 "Back-Transfer" from the Language of Mathematics to Applications

Next, the efficiency of the aforementioned exploration approach (Method 1) should be discussed for a standard test example: The *Marmousi model* (see [78, 116]) is a synthetic velocity data set which is well known and often used as a geological reference for density signature determination (see Fig. 3.14). It was first created by the Geophysics Group of the Institut Francais du Petrole in 1988. The Marmousi model is based on a profile through the North Quenguela Trough in the Cuanza Basin in Angola. Our interest is in the decorrelation of the geological signatures of this test area (actually, we use a canonically constructed 3D-version of the Marmousi model as used in [10]). In accordance with this standard test model, the contrast density function F is available as a fully interpreted 3D-Marmousi density model extension (see Fig. 3.13).

In order to validate the decorrelation abilities of our multiscale mollifier approach we use the de la Vallée Poussin-type mollifier multiscale method (cf. [39]) instead of the Haar approach, in order to have continuous singular integral kernels.

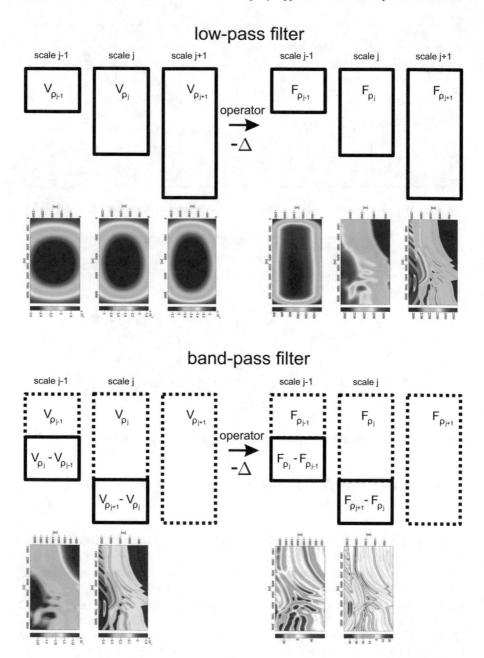

Fig. 3.15 Schematic visualization of the multiscale decorrelation mechanism in inverse gravimetry (see [10])

The results of our numerical experiences (see Fig. 3.15) may be summarized as follows: The low-pass filtered versions of the "Marmousi Newton potential" are very "smooth," so that the ϱ_j-potential functions V_{ϱ_j} provide no essential structural information for exploration purposes even for larger scales j. However, for smaller scale values ϱ_j, by going over to finer detail information involving ϱ_j-wavelet "Marmousi Newton potential functions," we already notice essential trends of the geological situation of the original density model. The reason is that the wavelet differences between potential function mollifiers may be regarded as some kinds of discretizations of the (negative) Laplace operator, i.e., the Poisson equation (3.7) becomes transparent in a discrete form.

The low-pass filtered density functions provide essential structural geological information, particularly for larger scales j. The band-pass (detail) density functions show strong separation surfaces between geological formations, which is of great significance in geothermal applications (where fracture transitions play a particular role for detecting aquifers and areas of internal water flow).

The results of the Marmousi example obtained by the PtJ-project SPE (Project Management Jülich, Federal Ministry for Economic Affairs and Energy, Berlin, funding reference number: 0324061, PI Prof. Dr. W. Freeden, CBM—Gesellschaft für Consulting, Business und Management mbH, Bexbach, Germany, corporate manager Prof. Dr. M. Bauer) demonstrate that the multiscale mollifier method of constructing (negative) Laplacian antiderivatives to singular integrals (such as Haar or de la Vallée Poussin-type singular integrals) is able to associate with geological formations certain wavelet band structures, so that geological interpretations specifically become obvious for purposes of exploration, thereby, e.g., reducing drastically the risk of an investor involved in geothermal power plant construction.

3.2 Circuit: Reflection Seismics

Next, once more based on a variant of mollifier regularization, wavelet techniques for the decorrelation, i.e., band-pass filtering of acoustic seismic phenomena, are discussed in order to get a local understanding and interpretability of scattered wave field potentials (our considerations essentially follow [27] and are based on the programmatic presentation [29]).

Background Knowledge All seismic methods in use can be distinguished between time- or depth-migration strategies and between application to post- or pre-stack data sets. The time-migration strategy is used to resolve conflicts in dipping events with different velocities. The depth-migration strategy handles strong lateral velocity variations associated with complex overburden structures.

The numerical techniques used to solve the migration problem can be separated generally into three broad categories:

- integral discretization methods such as Kirchhoff migration based on the solution of the eikonal equation,

- methods based on finite-difference schemes, e.g., depth continuation methods, reverse-time migration,
- transform methods based on frequency-wavenumber implementations, e.g., frequency-space and frequency-wavenumber migration.

All these migration methods usually rely on a certain approximation of the scalar acoustic or vectorial elastic wave equation (for more details the reader is referred, e.g., to [100] and the references therein).

According to the geophysical requirements, highly accurate approximations and efficient numerical techniques must be realized in order to handle steep dipping events and complex velocity models with strong lateral and vertical variations, as well as to construct the subsurface image in a locally defined region with high resolution on available computational resources. Consequently, migration algorithms require an accurate velocity model. The adaptation of the interval velocity by use of an inversion by comparing the measured travel times with simulated travel times is usually called reflection tomography. There are many versions of reflection tomography, but they all use ray-tracing techniques and they are formulated usually as a mathematical optimization problem. The most popular and efficient methods are ray-based travel time tomography, waveform- and full wave inversion (FWI) tomography , and Gaussian beam tomography (GBT) (see, e.g., [100, 123] and the references therein for more detailed information).

The "true" velocity estimation is often obtained by an iterative process called migration velocity analysis (MVA), which uses the kinematic information gained by the migration and consists of the following steps: (initial step) perform a reflection tomography of the coarse velocity structure using a priori knowledge about the subsurface; (iterative step) migrate the seismic data sets and apply the imaging condition; and update the velocity function by tomography inversion. These approaches often have rather poor quality with respect to the interpretability of the migration result. Since the interest of, e.g., geothermal projects is focused not only on structure heights and traps as in oil field practice, but also on fault zones and karst structures under recent stress conditions, the interpretation for geothermal needs is significantly more complicated.

3.2.1 "Transfer" from Applications to the Language of Mathematics

In the context of seismic processing, it is often assumed that shear stresses generated by the wave impulse and other kinds of damping can be neglected. As a consequence, wave propagation is treated as an acoustic phenomenon: A *pressure change* $P(x, t)$ dependent on the location x and the time t implies a volume change that generates a *displacement* $u(x, t)$ which yields further pressure changes in the neighborhood of the volume.

The relation between pressure changes P and volume changes dV is usually assumed to be governed by Hooke's law (see, e.g., [1, 111])

$$P = -K \frac{dV}{V} \tag{3.82}$$

as the basis of a linear elastic relation. Here, K is the *bulk modulus of the material*.

In order to connect the pressure change to the displacement (see also [42]), we formally observe that a small volume V may be written as $V = dx_1 dx_2 dx_3$. It is transformed to $V' = dx'_1 dx'_2 dx'_3$. If the displacement δu is given by

$$dx'_i = dx_i + \delta u_i, \quad i \in \{1, 2, 3\}, \tag{3.83}$$

our formal consideration leads to

$$
\begin{aligned}
-\frac{dV}{V} &= \frac{V - V'}{V} \\
&= \frac{dx_1 dx_2 dx_3 - (dx_1 + \delta u_1)(dx_2 + \delta u_2)(dx_3 + \delta u_3)}{dx_1 dx_2 dx_3} \\
&= -\frac{dx_1 dx_3 \delta u_2 + dx_2 dx_3 \delta u_1 + dx_1 dx_2 \delta u_3}{dx_1 dx_2 dx_3} \\
&\quad - \frac{dx_1 \delta u_2 \delta u_3 + dx_2 \delta u_1 \delta u_3 + dx_3 \delta u_1 \delta u_2}{dx_1 dx_2 dx_3} - \frac{\delta u_1 \delta u_2 \delta u_3}{dx_1 dx_2 dx_3} \\
&= -\left(\frac{\delta u_1}{dx_1} + \frac{\delta u_2}{dx_2} + \frac{\delta u_3}{dx_3} \right) + R \\
&= -\nabla \cdot u + R, \tag{3.84}
\end{aligned}
$$

where R collects all terms of higher order in terms of δu_i, so that R may be supposed to be negligible as δu_i is assumed to be small. Hence, we finally obtain

$$P = -K \nabla \cdot u + S \tag{3.85}$$

with a source term S as a constitutive equation. Moreover, we assume the balance of linear momentum that is equivalent to Newton's second law, which in our case may be formulated in the form

$$P(x + \delta x_i \varepsilon^i, t) - P(x, t) = -\varrho(x)\, \delta x_i a_i, \quad i \in \{1, 2, 3\}, \tag{3.86}$$

with the acceleration a and the canonical orthonormal basis of Euclidean space $\mathbb{R}^3$: $\varepsilon^1, \varepsilon^2, \varepsilon^3$. Under further assumption that the acceleration is given by the second order time derivative of the displacement u we get

$$\frac{P(x + \delta x_i \varepsilon^i, t) - P(x, t)}{\delta x_i} = -\varrho(x)\frac{\partial^2}{\partial t^2}u_i(x, t) \tag{3.87}$$

and finally by considering the limit $\delta x_i \to 0$ we end up with

$$\frac{\partial}{\partial x_i}P(x, t) = -\varrho(x)\frac{\partial^2}{\partial t^2}u_i(x, t). \tag{3.88}$$

These component equations can be expressed in vectorial form as follows:

$$\nabla_x P(x, t) = -\varrho(x)\frac{\partial^2}{\partial t^2}u(x, t). \tag{3.89}$$

Assuming that P and u are sufficiently often differentiable, the Eqs. (3.85) and (3.89) can be combined by applying the second order time derivative to (3.85) in the following way:

$$\frac{\partial^2}{\partial t^2}P(x, t) = -K(x)\,\nabla_x \cdot \left(\frac{\partial^2}{\partial t^2}u(x, t)\right) + \frac{\partial^2}{\partial t^2}S(x, t). \tag{3.90}$$

Inserting (3.89) into (3.90) we obtain

$$\frac{\partial^2}{\partial t^2}P(x, t) = K(x)\,\nabla_x \cdot \left(\frac{1}{\varrho(x)}\nabla_x P(x, t)\right) + \frac{\partial^2}{\partial t^2}S(x, t). \tag{3.91}$$

Applying the product rule yields the identity

$$\nabla_x \cdot \left(\frac{1}{\varrho(x)}\nabla_x P(x, t)\right) \tag{3.92}$$

$$= \frac{1}{\varrho(x)}\left(\underbrace{\nabla_x \cdot \nabla_x P(x, t)}_{=\Delta_x}\right)$$

$$- \frac{1}{\varrho^2(x)}\left(\nabla_x \varrho(x)\right) \cdot \left(\nabla_x P(x, t)\right), \tag{3.93}$$

provided that ϱ is smooth enough. If the gradient of ϱ is negligibly small, we arrive at the non-divergence form of the acoustic wave equation

$$\frac{\partial^2}{\partial t^2}P(x, t) = c^2(x)\,\Delta_x P(x, t) + \frac{\partial^2}{\partial t^2}S(x, t), \tag{3.94}$$

where the quantity

$$c(x) := \sqrt{\frac{K(x)}{\varrho(x)}} \tag{3.95}$$

is called the *propagation speed of a wave*, what results in a purely *divergence form of the acoustic wave equation*

$$\left(\frac{1}{c^2(x)}\frac{\partial^2}{\partial t^2} - \Delta_x\right) P(x,t) = \frac{1}{c^2(x)}\frac{\partial^2}{\partial t^2} S(x,t). \tag{3.96}$$

The identity (3.96) ends the standard approach to the acoustic wave equation, thereby assuming a compressible, viscous (i.e., no attenuation) medium with no shear strength and no internal forces (i.e., in equilibrium).

It should be mentioned that there is a large literature dealing with existence and uniqueness of the forward formulation of the acoustic wave equation (see, e.g., [20] and the references therein).

3.2.2 Mathematical Analysis

A standard approach to acoustic wave equation (3.96) is a Fourier transform with respect to time

$$U(x) = \frac{1}{\sqrt{2\pi}} \int_{\mathbb{R}} P(x,t) \exp(-i\omega t) \, dt \tag{3.97}$$

leading to the reduced wave equation

$$\left(\Delta_x + \frac{\omega^2}{c^2(x)}\right) U(x) = 0, \tag{3.98}$$

also called the *Helmholtz equation*.

In what follows, we are interested in two solution procedures of (3.98), namely

1. the decorrelation of U and c from a pre-processed (sufficiently suitable) solution U by use of an appropriate multiscale procedure (*postprocessing*),
2. the determination of c under the a priori knowledge of a low-pass filter, i.e., a "trend solution" within a multiscale procedure (*inverse modeling*).

In fact, for purposes of postprocessing, we do not make the attempt to solve the inverse problem of determining c as a whole. Instead we base our investigations on already successful pre-work, for example, (i) integral discretization methods such as Kirchhoff migration based on the solution of the eikonal equation (see, e.g., [123] and the references therein), Gaussian beam procedures (see, e.g., [103, 104]); (ii) methods based on the finite-difference schemes, e.g., depth continuation methods (cf. [13]), reverse-time migration (see, e.g., [7, 104, 123]); (iii) transform methods based on frequency-wavenumber implementations, e.g., frequency-space and frequency-wavenumber migration (see, e.g., [13, 123]). As a consequence, in order to obtain information about the structure, depth, and thickness of a target reservoir we start from the today's realistic assumption that standard seismic tomography results are available and meaningful, at least to some extent.

Altogether, postprocessing focuses our attention on the interpretability of a "true" migration result obtained from elsewhere and/or the (local) improvement of a trend result, and our specific mathematical merit merely is the postprocessing realization by using a new class of locally supported Helmholtz-type wavelets (based on the ideas of [29]) derived from the suitable "mollification" (regularization) of the fundamental solution of the Helmholtz operator.

3.2.3 Development of a Mathematical Solution Method

As alluded earlier, we first discuss the problem of postprocessing, then we go over to the problem of inversion.

Postprocessing The Helmholtz equation (3.98) leads to the definition of *the wave number $k(x)$* and the *refraction index $N(x)$* as

$$N(x) := \frac{c_0}{c(x)}, \quad k(x) := \frac{\omega}{c(x)} = \frac{\omega}{c_0} \frac{c_0}{c(x)} = k_0 N(x), \tag{3.99}$$

with c_0 being a suitable constant reference velocity and $k_0 := \frac{\omega}{c_0}$. Accordingly, the Helmholtz equation (3.98) can be rewritten in the form

$$\left(\Delta_x + k_0^2 N^2(x)\right) U(x) = 0. \tag{3.100}$$

The region where $N(x) \neq 1$ represents the scattering object such that $N(x) - 1$ may be supposed to set a compact support. Another standard assumption is that the difference between $c(x)$ and c_0 should be sufficiently small. As a consequence, $N^2(x)$ may be expanded into a Taylor series up to order one with a center such that $c(x_0) = c_0$. This yields the relation

$$N^2(x) \cong 1 + \epsilon v(x) \tag{3.101}$$

with a small perturbation parameter ϵ. Consequently, we have

$$k^2(x) := k_0^2 N^2(x) = k_0^2 \left(1 + \epsilon v(x)\right). \tag{3.102}$$

With the same argument as explained before, the unknown function $v(x)$ may be supposed to have a compact support. In accordance with the standard approach, the *wave operator*

$$A_x := \Delta_x + \frac{\omega^2}{c^2(x)} \tag{3.103}$$

may be separated in the following way:

$$
\begin{aligned}
A_x = \Delta_x + \frac{\omega^2}{c^2(x)} &= \Delta_x + k_0^2 N^2(x) = \Delta_x + k_0^2 \left(1 + \epsilon v(x)\right) \\
&= \Delta_x + k_0^2 + \epsilon k_0^2 v(x) = A_x^{(0)} + \epsilon A_x^{(1)},
\end{aligned} \tag{3.104}
$$

where we have used the abbreviations

$$A_x^{(0)} := \Delta_x + k_0^2 \tag{3.105}$$

and

$$A_x^{(1)} := k_0^2 v(x). \tag{3.106}$$

Hence, the *wave field U* may be split into an *incident wave field U_I*, corresponding to the wave propagating in the absence of the scatterer, and the *scattered wave field U_S*, such that

$$U = U_I + U_S. \tag{3.107}$$

This splitting leads to the equations

$$A^{(0)} U_I = \left(\Delta_x + k_0^2\right) U_I = 0, \tag{3.108}$$

$$A^{(0)} U_S = \left(\Delta_x + k_0^2\right) U_S = -\epsilon k_0^2 v(U_I + U_S) = -\epsilon k_0^2 v U = -\epsilon A^{(1)} U. \tag{3.109}$$

Once more, it should be mentioned that the Eq. (3.108) formalizes that U_I corresponds to the wave propagating in the absence of the scatterer.

As the fundamental solution to the Helmholtz operator $\Delta + k_0^2$ is known (up to a minus sign) to be

$$G(\Delta + k_0^2; |x - y|) := \frac{1}{4\pi} \frac{e^{ik_0|x-y|}}{|x - y|}, \quad x \neq y, \tag{3.110}$$

the functions U_I and U_S, respectively, can be represented as volume potentials

$$U_I(x) = \int_{\mathbb{R}^3} G(\Delta + k_0^2; |x - y|) \, W(y) \, dV(y), \tag{3.111}$$

$$U_S(x) = \int_{\mathcal{G}} G(\Delta + k_0^2; |x - y|) \underbrace{\left(-\epsilon k_0^2 \, v(y) \, U(y)\right)}_{=:F(y)} dV(y) \tag{3.112}$$

with the volume element dV and $\overline{\mathcal{G}} = \mathrm{supp}(v)$ being the local compact support of v, where W is given by

$$W(x) = -\frac{1}{\sqrt{2\pi}} \int_{\mathbb{R}} \frac{1}{c^2(x)} \frac{\partial^2}{\partial t^2} S(x, t) \exp(-i\omega t) \, dt. \tag{3.113}$$

It should be remarked that, in seismic reflection modeling (see Figs. 3.16, 3.17, 3.18, 3.19, and 3.20), point sources with certain spectra are usually chosen as unperturbed wave (for more details see, e.g., [100, 112]).

Only if U_I and U_S are available in the compact support $\overline{\mathcal{G}} = \mathrm{supp}(v)$, a direct computation of v by applying the Helmholtz operator $\Delta + k_0^2$ to the representation (3.112) becomes possible. However, it should be mentioned that U_S can be usually measured only away from the support $\overline{\mathcal{G}}$ of v. Nevertheless, in exceptional cases, local information is available inside boreholes, which is of particular interest in geothermal research.

In the sense of the perturbation theory and with the conventional setting $U^0 = U_I$, U can be formally written as a series

$$U = \sum_{k=0}^{\infty} \epsilon^k U^{(k)}, \tag{3.114}$$

which yields

$$\sum_{k=0}^{\infty} \epsilon^k \left(A^{(0)} + \epsilon A^{(1)}\right) U^{(k)} = W. \tag{3.115}$$

By collecting terms which are of the same order in ϵ we therefore get

$$A^{(0)} U^{(0)} = W, \tag{3.116}$$

$$A^{(0)} U^{(k)} = -A^{(1)} U^{(k-1)}, \quad k \in \mathbb{N}. \tag{3.117}$$

The scattered wave field is then given by

$$U_S = U - U_I = \sum_{k=1}^{\infty} \epsilon^k U^{(k)}. \tag{3.118}$$

Fig. 3.16 Wave propagation in the Marmousi velocity model after 0.6 s

Fig. 3.17 Wave propagation in the Marmousi velocity model after 1.1 s

Fig. 3.18 Wave propagation in the Marmousi velocity model after 1.6 s

Fig. 3.19 Wave propagation in the Marmousi velocity model after 2.1 s

Fig. 3.20 Wave propagation in the Marmousi velocity model after 2.6 s (from [68])

This procedure is known as *Born-approximation* (see, e.g., [112] for more details). Considering only the first order approximation, we obtain

$$A^{(0)}U_I = W, \tag{3.119}$$

$$A^{(0)}U_S = -\epsilon A^{(1)}U_I. \tag{3.120}$$

The difference between (3.109) and (3.120) is crucial. On the right-hand side of (3.112), we are confronted with the sum of U_I and U_S, which makes this a nonlinear equation. On the right-hand side of (3.120), only U_I appears, which is determined by (3.119), making the relation between scattered wave and perturbation of the medium linear.

Note that U_I and the first order approximation of U_S can be represented by the volume potentials

$$U_I(x) = \int_{\mathbb{R}^3} G(\Delta + k_0^2; |x-y|) \, W(y) \, dV(y), \tag{3.121}$$

$$U_S^I(x) = \int_{\mathcal{G}} G(\Delta + k_0^2; |x-y|) \underbrace{\left(-\epsilon k_0^2 \, v(y) \, U_I(y)\right)}_{=:F_I(y)} dV(y). \tag{3.122}$$

The basic properties of volume integrals of the type (3.112) can be summarized as follows (see, e.g., [90]): The volume potential U_S is a metaharmonic function in $\mathbb{R}^3 \backslash \overline{\mathcal{G}}$ under the assumption of boundedness of F_I (i.e., $(\Delta + k_0^2)U_S = 0$ in $\mathbb{R}^3 \backslash \mathcal{G}$). For a continuous F_I in $\overline{\mathcal{G}}$ the potential (3.112) is of class $C^{(1)}(\mathbb{R}^3)$ and, under the assumption of Hölder-continuity of F_I, we have (cf. [90])

$$(\Delta + k_0^2)U_S^I(x) = -F_I(x) = k_0^2 \, \epsilon v(x) \, U_I(x) \cong k_0^2 (N^2(x) - 1) \, U_I(x) \tag{3.123}$$

for all $x \in \mathcal{G}$. This equation indicates the direct relation between U_S^I and F_I in $\mathcal{G}$. It is actually the key point of Born modeling as discussed in the literature (see, e.g., [76]). Moreover, it parallels the approach already known for the Laplace operator in gravimetric obligations, which led to mollifier regularizations.

Multiscale Mollifier Approximation Our multiscale procedure starts from the nonlinear integral relation (3.111). Analogously to gravimetry, the idea is to use a sequence of *"mollifier" regularizations* $\{G_{\tau_j}(\Delta + k_0^2; \cdot)\}, j \in \mathbb{N}$, for the kernels (3.110) given by

$$G_{\tau_j}(\Delta + k_0^2; |x-y|) = \begin{cases} \frac{e^{ik_0|x-y|}}{4\pi|x-y|}, & |x-y| > \tau_j, \\ \frac{e^{ik_0|x-y|}}{8\pi\tau_j}\left(3 - \frac{|x-y|^2}{\tau_j^2}\right), & |x-y| \le \tau_j, \end{cases} \tag{3.124}$$

where $\{\tau_j\}_{j\in\mathbb{N}}$ denotes a positive monotonously decreasing sequence converging to 0 (for example, a dyadic sequence given by $\tau_j = 2^{-j}$, $j \in \mathbb{N}$). The regularization (3.124) is constructed in such a way that each kernel $G_{\tau_j}(\Delta + k_0^2; \cdot)$ is continuously differentiable and only dependent on the distance between two points x and y. Furthermore, under the assumption that F is continuous in $\overline{\mathcal{G}}$, we obtain for the "mollified version of the potential" (3.112)

$$(U_S)_{\tau_j}(x) = \int_{\mathcal{G}} G_{\tau_j}(\Delta + k_0^2; |x - y|) \underbrace{\left(-\epsilon k_0^2\, v(y)\, U(y)\right)}_{=F(y)} dV(y) \qquad (3.125)$$

the limit relation

$$U_S(x) = (U_S)_{\tau_j}(x) + O(\tau_j^2), \quad \tau_j \to 0, \qquad (3.126)$$

for all $x \in \mathbb{R}^3$. It should be noted that the aforementioned approach of replacing the fundamental solution $G(\Delta + k_0^2; |x - y|)$ by its regularized versions (3.125) initiates a multiscale method in already known way:

$$\lim_{j\to\infty} \sup_{x\in\overline{\mathcal{G}}} \left| U_S(x) - \int_{\mathcal{G}} G_{\tau_j}(\Delta + k_0^2; |x - y|) F(y)\, dV(y) \right| = 0. \qquad (3.127)$$

Each "scaling function" $G_{\tau_j}(\Delta + k_0^2; \cdot)$ provides low-pass filtering of the signature U_S. In order to obtain multiscale components, we again calculate the difference of two consecutive scaling functions and obtain the wavelet functions with respect to the scale parameter j. Of course, other types of regularizations can be chosen (as indicated in [39]). For simplicity, however, we again restrict ourselves to Haar-type kernel functions. By using the mollifications of the fundamental solution of the Helmholtz operator $\Delta + k_0^2$, we are immediately led to *locally supported Helmholtz-type wavelets* via the scale-discrete scaling equation

$$\Psi_{\tau_j}(\Delta + k_0^2; |x - y|) = G_{\tau_{j+1}}(\Delta + k_0^2; |x - y|) - G_{\tau_j}(\Delta + k_0^2; |x - y|), \quad j \in \mathbb{N}_0. \qquad (3.128)$$

Explicitly, we have (see Fig. 3.21 for a graphical illustration)

$$\Psi_{\tau_j}(\Delta + k_0^2; |x - y|) = \qquad\qquad\qquad\qquad\qquad\qquad (3.129)$$

$$\begin{cases} \frac{e^{ik_0|x-y|}}{8\pi\,\tau_{j+1}}\left(3 - \frac{|x-y|^2}{\tau_{j+1}^2}\right) - \frac{e^{ik_0|x-y|}}{8\pi\,\tau_j}\left(3 - \frac{|x-y|^2}{\tau_j^2}\right), & |x - y| \leq \tau_{j+1}, \\[2ex] -\frac{e^{ik_0|x-y|}}{4\pi|x-y|} + \frac{e^{ik_0|x-y|}}{8\pi\,\tau_j}\left(3 - \frac{|x-y|^2}{\tau_j^2}\right), & \tau_{j+1} \leq |x - y| \leq \tau_j, \\[2ex] 0, & \tau_j \leq |x - y| \ . \end{cases}$$

The convolution integral (3.130) indicates the difference of two sequential low-pass filters, i.e., it represents a band-pass filtering at the position x with respect to the

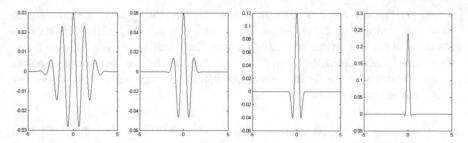

Fig. 3.21 Wavelet function Ψ_{τ_j} in sectional illustration for $k_0 = 5$ and $\tau_j = \frac{4}{2^j}$, $j = 0, \ldots, 3$

scale parameter j

$$W_{\tau_j}(x) = \int_{\mathcal{G}} \Psi_{\tau_j}(\Delta + k_0^2; |x - y|) \, F(y) \, dV(y). \tag{3.130}$$

W_{τ_j} includes all detail information contained in $(U_S)_{\tau_{j+1}}$ but not in $(U_S)_{\tau_j}$. In accordance with our construction we therefore obtain for every $L \in \mathbb{N}$

$$(U_S)_{\tau_{j+L}} = (U_S)_{\tau_j} + \sum_{n=j}^{j+L-1} W_{\tau_n}. \tag{3.131}$$

The formula (3.131) shows the progress in the "zooming-in process," which proceeds from scale j to scale $j + L$. Hence, the identity (3.131) describes the amount of improvement in the accuracy from level j to level $j + L$. Indeed, it uniformly follows for each position x and for each scale value j that

$$U_S(x) = (U_S)_{\tau_n}(x) + \sum_{n=j}^{\infty} W_{\tau_n}(x), \tag{3.132}$$

i.e., the signal U_S consists of a (coarse) low-pass filtering and an infinite number of successive band-pass convolutions. Of course, in practice, only a finite number of band-pass filters have to be calculated to satisfy a certain error tolerance.

Until now, we have only reconstructed the quantity U_S by virtue of a multiscale technique using building blocks. For practical purposes, the decorrelation of both $F = -\epsilon \nu \, k_0^2 U$ and $-\epsilon \nu \cong 1 - N^2$ (cf. (3.101)) is of interest. An elementary calculation using (3.124) yields the Helmholtz derivative

$$K_{\tau_j}(\Delta + k_0^2; |x - y|) = -\left(\Delta_x + k_0^2\right) G_{\tau_j}\left(\Delta + k_0^2; |x - y|\right)$$

$$= \begin{cases} -3\dfrac{e^{ik_0|x-y|}}{4\pi \, |x-y| \, \tau_j^3}(ik_0\tau_j^2 - ik_0|x - y|^2 - |x - y|) \,, & |x - y| \le \tau_j, \\ 0 & , \ |x - y| > \tau_j. \end{cases}$$

$\{K_{\tau_j}\}$ is a "Haar-type sequence," and it approximately reduces to the Haar sequence $\{H_{\tau_j}\}$

$$H_{\tau_j}(|x - y|)) = \begin{cases} 3\frac{1}{4\pi\tau_j^3} & , \ |x - y| \le \tau_j, \\ 0 & , \ |x - y| > \tau_j. \end{cases}$$

for sufficiently large j, i.e., for each fixed k_0

$$K_{\tau_j}(\Delta + k_0^2; |x - y|) = H_{\tau_j}(|x - y|) + O(\tau_j), \quad j \to \infty. \tag{3.133}$$

If F is continuous, then it is clear that

$$\left(\Delta_x + k_0^2\right) \int_{\mathcal{G}} G_{\tau_j}(\Delta + k_0^2; |x - y|) \, F(y) \, dV(y) \tag{3.134}$$

$$= \int_{\mathcal{G}} \left(\Delta_x + k_0^2\right) G_{\tau_j}\left(\Delta + k_0^2; |x - y|\right) F(y) \, dV(y).$$

and

$$F(x) = \lim_{j \to \infty} \int_{\mathcal{G}} K_{\tau_j}(\Delta + k_0^2; |x - y|) \, F(y) \, dV(y)$$

$$= \lim_{j \to \infty} \int_{\mathcal{G}} H_{\tau_j}(|x - y|) \, F(y) \, dV(y) \tag{3.135}$$

for all $x \in \mathcal{G}$. Moreover, we have under the assumption of Hölder-continuity of F (see [90])

$$F(x) = -(\Delta_x + k_0^2) \, U_S(x) = -(\Delta_x + k_0^2) \int_{\mathcal{G}} G(\Delta + k_0^2; |x - y|) F(y) \, dV(y) \tag{3.136}$$

for all $x \in \mathcal{G}$. In other words, the negative "Helmholtz derivative" $-(\Delta + k_0^2)$ of the mollifier regularization of the fundamental solution leads back to a "Haar-type" singular integral (3.135) for detecting

$$F(x) = -k_0^2 \, \epsilon v(x) \, (U_I + U_S)(x) \cong k_0^2 \, (1 - N^2(x)) \, U(x) \tag{3.137}$$

for all $x \in \mathcal{G}$.

Finally it should be once more pointed out that our approach also offers the possibility to introduce alternative mollifier regularizations of the fundamental solution (cf. [41]) such that their (negative) "Helmholtz derivatives" represent singular-type kernels constituting Dirac-type sequences in $\mathcal{G}$. In all these cases, our multiscale method by mollified fundamental solutions enables us to decompose simultaneously the signal information of the wave field as well as the refraction index N based on the interrelation (3.133); however, under the "postprocessing assumption" that U_S is discretely known (to some extent) inside $\mathcal{G}$. Once more, our understanding of the multiscale technique here does not preferably aim at the reconstruction of the signals, but instead in working out characteristic detail information that emerge from the difference of two consecutive scale-space representations. This practice comes across as the decorrelation of signatures, such that the postprocessing of certain band structures showing typical features for exploration purposes is the key element.

Mollifier Inversion Next we apply the Haar philosophy to indicate an approximate determination of F from U_S inside $\mathcal{G}$. The aforementioned mollifier regularization procedure of the volume potential (as proposed for postprocessing) is the essential tool. Let $\{\tau_j\}_{j \in \mathbb{N}_0}$ be a monotonously decreasing sequence of positive values τ_j such that $\lim_{j \to \infty} \tau_j = 0$ (for example, $\tau_j = 2^{-j}$). Then, in accordance with (3.135), we are able to specify a sufficiently large integer J such that, for all $x \in \mathcal{G}$,

$$F(x) \cong F_{\tau_J}(x) = \int_{\mathcal{G}} K_{\tau_J}(\Delta + k_0^2; |x - y|)\, F(y)\, dV(y) \tag{3.138}$$

as well as

$$(U_S)(x) \cong (U_S)_{\tau_J}(x) = \int_{\mathcal{G}} G_{\tau_J}(\Delta + k_0^2; |x - y|)\, F(y)\, dV(y) \tag{3.139}$$

(as always, "$\cong$"means that the error is negligible). If F is continuous, then we already know that

$$-\left(\Delta_x + k_0^2\right)(U_S)_{\tau_J}(x) = F_{\tau_J}(x) \tag{3.140}$$

for all $x \in \mathcal{G}$. In order to realize a fully discrete approximation of F, we have to apply approximate integration formulas over $\overline{\mathcal{G}}$ leading to

$$(U_S)(x) \cong \sum_{i=1}^{N_J} \underbrace{w_i^{N_J} F(y_i^{N_J})}_{=a_i^{N_J}}\, G_{\tau_J}(\Delta + k_0^2; ; |x - y_i^{N_J}|), \tag{3.141}$$

where $w_i^{N_J}$, $y_i^{N_J}$, $i = 1, \ldots, N_J$, are the known weights and knots, respectively. Using an appropriate integration formula, we are therefore led to a linear system to be solved in order to obtain insight into approximate information about ν. For numerical realization we assume that $(U_S)_{\tau_{J-1}}$ is available, at least discretely in points $x_k^{M_J} \in \overline{\mathcal{G}}$, $k = 1, \ldots, M_J$, to be needed in the solution process of the linear system.

Obviously, we have to calculate all unknown coefficients

$$a_i^{N_J} = w_i^{N_J} F(y_i^{N_J}), \quad i = 1, \ldots, N_J \tag{3.142}$$

from discrete values $U_S\left(x_k^{M_J}\right) - (U_S)_{\tau_{J-1}}\left(x_k^{M_J}\right), k = 1, \ldots, M_J$. Then we have to solve a linear system

$$U_S\left(x_k^{M_J}\right) - (U_S)_{\tau_{J-1}}\left(x_k^{M_J}\right) = \sum_{i=1}^{N_J} \Psi_{\tau_{J-1}}(\Delta + k_0^2; |x_k^{M_J} - y_i^{N_J}|) \, a_i^{N_J}, \tag{3.143}$$

$k = 1, \ldots, M_J$, to determine the required coefficients $a_i^{N_J}, i = 1, \ldots, N_J$.

Once all values $F(y_i^{N_J}), i = 1, \ldots, N_J$, are available (note that the integration weights $w_i^{N_J}, i = 1, \ldots, N_J$, are known), the function F can be obtained from the formula

$$F(x) \cong F_{\tau_J}(x) = \sum_{i=1}^{N_J} K_{\tau_J}(\Delta + k_0^2; |x - y_i^{N_J}|) \underbrace{w_i^{N_J} F(y_i^{N_J})}_{=a_i^{N_J}} \tag{3.144}$$

for all $x \in \mathcal{G}$. In connection with the knowledge of $U_S, (U_S)_{\tau_{J-1}}$, we are therefore able to determine v, hence, the refraction index N via (3.137).

Clearly, the linear system (3.143) is a bottleneck in inverse modeling. The advantage of our construction principle, however, is that sparse solution techniques become applicable, since the *"Helmholtz-type wavelets"* $\Psi_{\tau_{J-1}}$ possess a local support that becomes smaller as J increases.

All in all, Helmholtz-type wavelets turn out to be not only numerically economic and efficient, but also close to the physical reality (as mollifier regularizations of the fundamental solution of the Helmholtz operator). In conclusion, Helmholtz-type wavelets as proposed here form a promising compromise in physics and numerics of acoustically based reflection seismics.

3.2.4 *"Back-Transfer" from the Language of Mathematics to Applications*

For test investigations we use the Marmousi migration model (see [77]) as shown in Fig. 3.22 (taken from [68]). The source wave field involving the Marmousi model in direct time (i.e., snapshots of the wave propagation after 0.6, 1.1, 1.6, 2.1, and 2.6 s) is illustrated in Figs. 3.16, 3.17, 3.18, 3.19, 3.20 (taken from [68]). In analogy to

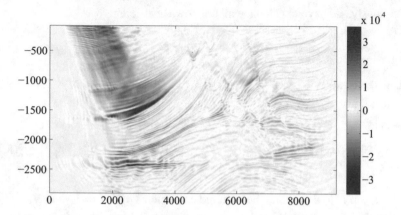

Fig. 3.22 A migration result of the Marmousi model by FDS (as used in [68])

inverse gravimetry, a schematic result applied to the Marmousi data set is illustrated in Fig. 3.23.

The results of our numerical experiences (in Fig. 3.23) obtained by the PtJ-project GEOFÜND (Project Management Jülich, Federal Ministry for Economic Affairs and Energy, Berlin, funding reference number: 0325512A, PI Prof. Dr. W. Freeden, University of Kaiserslautern, Germany) are quite similar to the case of inverse gravimetry: The low-pass filtered versions of the "Marmousi potential model" are "smooth," so that the τ_j-potential functions $(U_S)_{\tau_j}$ provide no essential structural information for exploration purposes even for larger scales j. In addition, they are superimposed by undesired noise phenomena caused by an erroneous migration inherently occurring in Fig. 3.22. By wavelet filtering the migration result, however, we are able to dampen undesired noise. In addition, for smaller scale values τ_j, by going over to finer detail information involving τ_j-wavelet "Marmousi potential functions" $(U_S)_{\tau_j}$, we already notice more essential trends reflecting the geology. The reason again is that the wavelet differences between potential function mollifiers may be regarded as some kinds of discretizations of the (negative) Helmholtz operator, i.e., the differential equation (3.136) becomes transparent in a discrete form. The band-pass (detail) functions $F_{\tau_{j+1}} - F_{\tau_j}$ show a strong scale τ_j-dependent separation into geologically reflected features. Moreover, these illustrations show marginally visible influence of noise. They are, indeed, very useful for a band-pass characterization of the refraction index via (3.101).

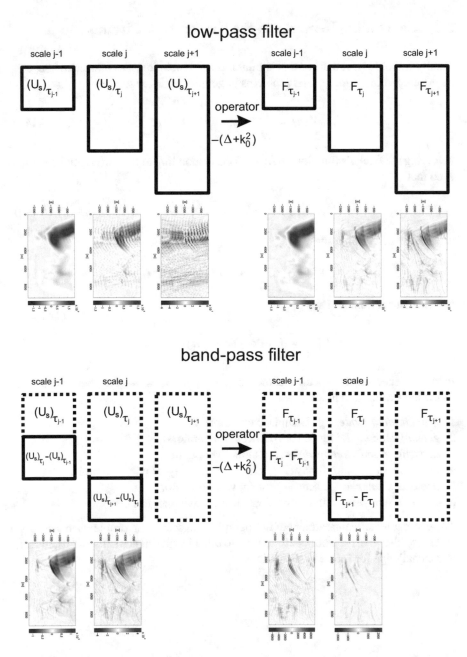

Fig. 3.23 Schematic visualization of the multiscale decorrelation mechanism in reflection seismics (see [10])

3.3 Concluding Remarks to the Exploration Circuits

Finally it should be remarked that similar mollifier techniques as presented here can be developed for more general problems in geomathematics such as

$$V(x) = \int_{\mathcal{G}} G(L; x, y) \, F(y) \, dy \tag{3.145}$$

involving the Green's function $G(L; x, y)$ corresponding to the differential operator L so that

$$F(x) = -L_x V(x), \ x \in \mathcal{G}, \tag{3.146}$$

where in distributional sense

$$- L_x G(L; x, y) = \delta(x, y), \ x \in \mathcal{G} \tag{3.147}$$

and (in distributional sense)

$$F(x) = \int_{\mathcal{G}} \delta(x, y) \, F(y) \, dy, \ x \in \mathcal{G}. \tag{3.148}$$

In dependence on the specific choice of the operator L we are then led to the following exploration areas:

1. *gravitational modeling* (L Laplace operator),
2. *geomagnetic modeling* (L (pre-) Maxwell operators),
3. *acoustic seismic tomography* (L Helmholtz operator),
4. *elastic seismic tomography* (L (Helmholtz) Cauchy–Navier operator),
5. *acoustic scattering* (L time dependent wave operator),
6. *elastic scattering* (L time dependent elastic wave operator).

The geomathematical treatment of tomography and scattering problems in a consistent setup also gives scientific information to other Earth science subsystems, in fact, it is a challenge for future investigations.

Chapter 4
Recent Activities in Geomathematics

The special importance of mathematics as an interdisciplinary science has been acknowledged increasingly in engineering, technology, economy, and commerce. This process does not remain without effects on mathematics itself. The mathematical "eye for similarities" created innovative solution methods and structures arising from very different areas and distinct situations, so that the resulting theories and models may be applicable to multiple situations after appropriate adaptation and concretization.

Today we notice several activities in the field of *geomathematics*:

- The last 20 years have been marked for various mathematical activities under a wide umbrella of initiatives called the Mathematics of the Planet Earth (MPE) 2013, which focused on mathematical research in areas of relevance to the various processes that affect the planet Earth. The dynamics of the oceans and the atmosphere and the changes in the climate are of course the obvious topics that are very important for the life on planet Earth and make use of mathematics in an essential way. In addition to these, a multitude of other topics are of relevance to MPE-2013, including the financial and economic systems, the energy production and utilization, spread of epidemics at the population level, ecology and genomics of species, just to name a few. To stimulate imagination on the many domains where mathematics plays a crucial role in planetary issues, the following four (nonexhaustive) themes are proposed as part of MPE-2013:

  1. A planet to discover: oceans; meteorology and climate; mantle processes, natural resources, celestial mechanics.
  2. A planet supporting life: ecology, biodiversity, evolution.

© The Author(s), under exclusive license to Springer Nature Switzerland AG 2019
W. Freeden et al., *An Invitation to Geomathematics*, Lecture Notes in Geosystems
Mathematics and Computing, https://doi.org/10.1007/978-3-030-13054-1_4

3. A planet organized by humans: political, economic, social and financial systems; organization of transport and communications networks; management of resources; energy.
4. A planet at risk: climate change, sustainable development, epidemics; invasive species, natural disasters.

Another major issue of the MPE initiative is public outreach to increase awareness of the importance and essential nature of mathematics in tackling these problems, and to bring out the relevance and usefulness of mathematics to a wider section of society than just those who use it professionally (see, e.g., https://www.icts.res.in/program/MPE2013 for more details).

- Mathematics of Planet Earth has highlighted the need for researchers with a broad view of planetary issues: climate change, quantification of uncertainty, move to an economy of sustainability, preservation of biodiversity, and adaptation to change. Joining forces with other disciplines and recruiting young researchers to work on these issues is a priority. This is why the *S*ociety of *I*ndustrial and *A*pplied *M*athematics (SIAM) has created SIAM Activity Groups on "Geosciences and Mathematics of Planet Earth." In this respect it should be noted that the term "geomathematics" was coined more than two decades ago, it may be used as the canonical abbreviation of both expressions "mathematical geosciences" and "mathematics of planet Earth."
- SIAM conferences focus on timely topics in modeling and simulation and their application in various fields of the geosciences and provide a place to exchange ideas and to expand their network of colleagues in both academia and industry. For example, SIAM GS 2017 included renewable energies (e.g., thermal, wind-driven), underground waste disposal and cleanup of hazardous waste, earthquake prediction, along with the well-established fields of petroleum exploration and recovery. As a consequence, SIAM conferences ensure the dissemination of appropriate tools and methods, and foster useful fundamental research in applied mathematics. These kinds of interactions are needed for meaningful progress in understanding and predicting complex physical phenomena in the geosciences.
- There are several geomathematical graduate programs in the USA, Germany, and other countries.
- A series of universitary chairs for geomathematics were founded. In Germany, it started with Kaiserslautern in the year 1994, Siegen and Freiberg followed in 2008 and 2018, respectively.

Appendix A
GEM International Journal
on Geomathematics

As already pointed out, the special importance of mathematics as a basic science that makes important contributions in technology, economy, and commerce has been increasingly acknowledged within the last few years. This process did not remain without effects on mathematics itself. New mathematical disciplines, such as scientific computing, financial and business mathematics, industrial mathematics, biomathematics, and also mathematics concerned with geoscientific problems, i.e., *geomathematics*, have complemented the traditional mathematical disciplines.

Mathematics is the "raw material" for the models and the essence of each computer simulation. As the key technology, it translates images of the real world to models of the virtual world, and vice versa.

GEM—International Journal on Geomathematics, ISSN: 1869–2672 (print version), ISSN: 1869–2680 (electronic version), https://www.springer.com/mathematics/applications/journal/13137, was launched in August 2010 as a forum for peer-reviewed mathematical papers that

1. model the system Earth (geosphere, cryosphere, hydrosphere, atmosphere, biosphere), or
2. contain analytic, algebraic, and operator-theoretic methods necessary for the mathematical treatment of geoscientifically relevant problems, or
3. apply geospecific computational and numerical analysis methods, or
4. represent survey contributions characterizing the bridge between a geoscientifically relevant research field and mathematics.

A.1 GEM and Its Task and Objective

The intent of the mathematical journal GEM is to deal with the qualitative and quantitative properties of the current or possible structures of the system Earth. GEM provides concepts of scientific research concerning the *system Earth*. The research

© The Author(s), under exclusive license to Springer Nature Switzerland AG 2019   93
W. Freeden et al., *An Invitation to Geomathematics*, Lecture Notes in Geosystems
Mathematics and Computing, https://doi.org/10.1007/978-3-030-13054-1

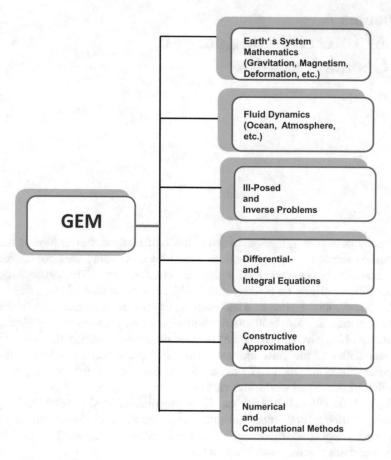

Fig. A.1 Essential thematic structure of GEM

object (cf. Fig. A.1), i.e., the system Earth, consists of a number of elements which represent individual systems themselves. The complexity of the entire system Earth is determined by interacting physical, biological, and chemical processes transforming and transporting energy, material, and information. It is characterized by natural, social, and economic processes influencing one another. All these aspects require a geoscientifically relevant type of mathematics. Geomathematics and its organ GEM have to be concerned with nothing more than the organization of the complexity of the system Earth. Descriptive thinking is required in order to clarify abstract complex situations. Correct simplifications of complicated interactions, exact thinking and formulations are needed, numerical realizations (modeling and simulation) should be given, etc. All in all, geomathematics represents the key science of the complex system Earth. Wherever there are data and observations to be processed, e.g., the diverse scalar, vectorial, and tensorial clusters of satellite and exploration data, we need geomathematics. Whenever modeling and simulation come into play for the system Earth, geomathematics is decisive.

A.2   GEM As Scientific Bridge

The specific task of geomathematics and its forum GEM is to build a bridge between mathematical theory and geophysical as well as geotechnical practice. The special attraction is based on the vivid communication between mathematicians more interested in model development, theoretical foundation, and the approximate as well as computational solution of problems, and engineers and physicists more familiar with measuring technology, methods of data analysis, implementation of routines, and software applications.

A special feature is that geomathematics primarily deals with those regions of the Earth which are only insufficiently or not at all accessible for direct measurements (even by remote sensing methods). Inverse theories and methods are absolutely essential for the mathematical evaluation in these cases. Mostly, a physical field is measured in the vicinity of the Earth's surface and/or satellite height, and it has then to be continued downward or upward by mathematical methods until one reaches the interesting depths or heights. For all these aspects GEM provides an adequate forum for scientific distribution among mathematically oriented people.

A.3   GEM and Its Editorial Structure

GEM is intended to publish peer-reviewed mathematical papers and survey articles of geoscientific relevance. The relevance does not include necessarily numerical applications to a real data scenario. Instead all mathematically applicable, i.e., theoretically and/or computationally significant facets and obligations of the Earth's system are welcome for scientific distribution by means of GEM.

In 2017, GEM essentially dealt with the following scientific topics and research areas (cf. Fig. A.1) associated with selected editors:

1. modeling and simulation problems of the system Earth (geosphere, cryosphere, hydrosphere, atmosphere, biosphere, anthroposphere),
2. fluid dynamics problems (e.g., ocean currents, atmospheric circulation, etc.),
3. inverse problems (e.g., in satellite technology, geoexploration, etc.),
4. geoscientifically relevant differential and integral equations (concerning systems of potential fields, diffusion fields, wave fields or combinations of them),
5. constructive approximation,
6. numerical methods, efficient and economical computational algorithms and procedures.

In addition, the role of GEM as a scientic bridge between geomathematics and other geoscientic disciplines plays a particular, whose articulation is found in survey papers by distinguished geoscientists.

Accordingly, in adaptation to the aforementioned list, future editorial work within the GEM-reviewing process will provide a more specific and specialized basis for collecting publishable contributions in certain areas of geomathematics.

All in all, the essential goal of the Editor-in-Chief as well as the publisher is that GEM more and more emerges as an active and dynamic journal for the dissemination of creative and innovative ideas and concepts with the motivation to help the mathematically interested readership to understand the Earth as our living environment. As in the last ten years, high mathematical quality in combination with georelevance will be canonically expected from the contributions to the journal.

Survey-like articles bridging between mathematical theory and geophysical as well as geotechnical practice constitute complementing research-oriented GEM areas for publication.

A.4   Previous Publications in GEM International Journal on Geomathematics

Since its launch the journal has published every year between 300 and 400 pages and has established itself as one of the leading journals in its field.

Vol. 1, Issue 1

GEM: International Journal on Geomathematics
Author: Willi Freeden

The 'Regiments' of sun and pole star: on declination tables in early modern England
Author: Thomas Sonar

Randomized anisotropic transform for nonlinear dimensionality reduction
Authors: Charles K. Chui, Jianzhong Wang

Limit formulae and jump relations of potential theory in Sobolev spaces
Authors: Martin Grothaus, Thomas Raskop

Regularisation of the Helmholtz decomposition and its application to geomagnetic field modelling
Authors: M. Akram, V. Michel

Estimation of linear functionals from indirect noisy data without knowledge of the noise level
Authors: Sergei V. Pereverzev, Bernd Hofmann

Vol. 1, Issue 2

Delayed progress in navigation: the introduction of line of position navigation in Germany and Austria
Author: Günther Oestmann

Space gradiometry: tensor-valued ellipsoidal harmonics, the datum problem and application of the Lusternik–Schnirelmann category to construct a minimum atlas
Author: Erik W. Grafarend

Spline multiresolution and numerical results for joint gravitation and normal-mode inversion with an outlook on sparse regularisation
Authors: Paula Berkel, Doreen Fischer, Volker Michel

Spherical decompositions in a global and local framework: theory and an application to geomagnetic modeling
Author: C. Gerhards

Modeling anomalous heat transport in geothermal reservoirs via fractional diffusion equations
Authors: Yury Luchko, Alessandro Punzi

Vol. 2, Issue 1

Hybrid regularization methods for seismic reflectivity inversion
Authors: Yanfei Wang, Yan Cui, Changchun Yang

Spatiospectral concentration in the Cartesian plane
Authors: Frederik J. Simons, Dong V. Wang

Three-dimensional modeling of heat transport in deep hydrothermal reservoirs
Author: Isabel Ostermann

Spherical fast multiscale approximation by locally compact orthogonal wavelets
Authors: Frank Bauer, Martin Gutting

A minimal atlas for the rotation group SO(3)
Authors: Erik W. Grafarend, Wolfgang Kühnel

Vol. 2, Issue 2

Natural convection in horizontal annuli: evaluation of the error for two approximations
Authors: Agnes Lamacz, Arianna Passerini, Gudrun Thäter

On homogenization of stokes flow in slowly varying media with applications to fluid–structure interaction
Authors: Donald L. Brown, Peter Popov, Yalchin Efendiev

Proof of validity of first-order seismic traveltime estimates
Authors: Len Bos, Michael A.

Explaining the seismic moment of large earthquakes by heavy and extremely heavy tailed models
Author: Miguel Martins Felgueiras

Fast multipole accelerated solution of the oblique derivative boundary value problem
Author: Martin Gutting

On instabilities in data assimilation algorithms
Authors: A. Boris, R. Marx, W. E. Potthast

Spherical decomposition of electromagnetic fields generated by quasi-static currents
Authors: Jin Sun, Gary D. Egbert

The geomagnetic field gradient tensor
Authors: Stavros Kotsiaros, Nils Olsen

Vol. 4, Issue 1

Visualising geomagnetic data by means of corresponding observations
Authors: Karin Reich, Elena Roussanova

A mathematical model of a passive scheme for acid mine drainage remediation
Authors: M. Gouin, E. Saracusa, C. B. Clemons, J. Senko, K. L. Kreider, G. W. Young

On the relationship between ray theory and the banana-doughnut formulation
Authors: Len Bos, Michael A. Slawinski

Different radial basis functions and their applicability for regional gravity field representation on the sphere
Authors: Katrin Bentel, Michael Schmidt, Christian Gerlach

A linear model for the sea breeze circulation relevant for the tropical regions
Authors: Arnab Jyoti Das Gupta, A. S. Vasudeva Murthy, Ravi S. Nanjundiah, C. V. Srinivas

Vol. 4, Issue 2

Wavelet packets for time-frequency analysis of multispectral imagery
Authors: John J. Benedetto, Wojciech Czaja, Martin Ehler

On the size and shape of drumlins
Authors: A. C. Fowler, M. Spagnolo, C. D. Clark, C. R. Stokes, A. L. C. Hughes, P. Dunlop

Meshfree generalized finite difference methods in soil mechanics—part I: theory
Authors: I. Ostermann, J. Kuhnert, D. Kolymbas, C.-H. Chen, I. Polymerou, V. Šmilauer, C. Vrettos, D. Chen

An improved version of a high accuracy surface modeling method
Authors: Na Zhao, Tianxiang Yue, Mingwei Zhao

Statistically-based approach for monitoring of micro-seismic events
Authors: A. Kushnir, N. Rozhkov, A. Varypaev

Exploration of a simple model for ice ages
Authors: A. C. Fowler, R. E. M. Rickaby, E. W. Wolff

Vol. 5, Issue 1

Compensating operators and stable backward in time marching in nonlinear
    parabolic equations
Author: Alfred S. Carasso

A recursive linear MMSE filter for dynamic systems with unknown state vector
    means
Authors: Amir Khodabandeh, Peter J. G. Teunissen

On the regular decomposition of the inverse gravimetric problem in non- L2 spaces
Author: Fernando Sansó

A multiscale power spectrum for the analysis of the lithospheric magnetic field
Author: Christian Gerhards

Regularized collocation for spherical harmonics gravitational field modeling
Valeriya Naumova, Sergei V. Pereverzyev, Pavlo Tkachenko

Potential modeling: conditional independence matters
Author: Helmut Schaeben

Decomposition of optical flow on the sphere
Authors: Clemens Kirisits, Lukas F. Lang, Otmar Scherzer

Vol. 5, Issue 2

Local numerical integration on the sphere
Authors: J. Beckmann, H. N. Mhaskar, J. Prestin

The adjoint method in geodynamics: derivation from a general operator formulation
    and application to the initial condition problem in a high resolution mantle
    circulation model
Authors: André Horbach, Hans-Peter Bunge, Jens Oeser

A non-linear approximation method on the sphere
Authors: Volker Michel, Roger Telschow

Generalized multiscale finite element method for elasticity equations
Authors: Eric T. Chung, Yalchin Efendiev, Shubin Fu

Time-space adaptive discontinuous Galerkin method for advection-diffusion equa-
    tions with non-linear reaction mechanism
Authors: Bülent Karasözen, Murat Uzunca

Estimating the number and locations of Euler poles
Authors: F. Bachmann, P. E. Jupp, H. Schaeben

Vol. 6, Issue 1

Signal analysis via instantaneous frequency estimation of signal components
Authors: Charles K. Chui, Maria D. van der Walt

Sampling, splines and frames on compact manifolds
Author: Isaac Z. Pesenson

Restoring past mantle convection structure through fluid dynamic inverse theory:
  regularisation through surface velocity boundary conditions
Authors: Lyudmyla Vynnytska, Hans-Peter Bunge

The reference figure of the rotating earth in geometry and gravity space and an
  attempt to generalize the celebrated Runge–Walsh approximation theorem for
  irregular surfaces
Author: Erik W. Grafarend

Vol. 6, Issue 2

Hierarchical multiscale modeling for flows in fractured media using generalized
  multiscale finite element method
Authors: Yalchin Efendiev, Seong Lee, Guanglian Li, un YaoNa Zhang

Multi-variate Hardy-type lattice point summation and Shannon-type sampling
Authors: Willi Freeden, M. Zuhair Nashed

Compression approaches for the regularized solutions of linear systems from large-
  scale inverse problems
Authors: Sergey Voronin, Dylan Mikesell, Guust Nolet

Rosborough approach for the determination of regional time variability of the
  gravity field from satellite gradiometry data
Authors: Wolfgang Keller, Rey-Jer You

Fourth order Taylor–Kármán structured covariance tensor for gravity gradient
  predictions by means of the Hankel transformation
Authors: Erik W. Grafarend, Rey-Jer You

Modeling erosion and sediment delivery from unpaved roads in the north mountain-
  ous forest of Iran
Authors: Abolfazl Jaafari, Akbar Najafi, Javad RezaeianAli Sattarian

Vol. 7, Issue 1

The compressible adjoint equations in geodynamics: derivation and numerical
  assessment
Authors: Siavash Ghelichkhan, Hans-Peter Bunge

A new hierarchically-structured n-dimensional covariant form of rotating equations
  of geophysical fluid dynamics
Author: Werner Bauer

Vol. 8, Issue 2

Friedrich Robert Helmert in memory of his 100th year of death
Author: Bertold Witte

Romberg extrapolation for Euler summation-based cubature on regular regions
Authors: Willi Freeden, Christian Gerhards

On the convergence theorem for the regularized functional matching pursuit
  (RFMP) algorithm
Authors: Volker Michel, Sarah Orzlowski

Meshfree generalized finite difference methods in soil mechanics—part II: numeri-
  cal results
Authors: I. Michel, S. M. I. Bathaeian, J. Kuhnert, D. Kolymbas, C.-H. Chen,
  I. Polymerou, C. Vrettos, A. Becker

Conjugate gradient based acceleration for inverse problems
Authors: Sergey Voronin, Christophe Zaroli, Naresh P. Cuntoor

Single-species model under seasonal succession alternating between Gompertz and
  Logistic growth and impulsive perturbations
Authors: Yanqing Li, Long Zhang, Zhidong Teng

Vol. 9, Issue 1

Operator-theoretic and regularization approaches to ill-posed problems
Authors: Willi Freeden, M. Zuhair Nashed

Monte Carlo methods
Author: Karl-Rudolf Koch

The inverse scattering problem for orthotropic media in polarization-sensitive
  optical coherence tomography
Authors: Peter Elbau, Leonidas Mindrinos, Otmar Scherzer

Vol. 9, Issue 2

A greedy algorithm for nonlinear inverse problems with an application to nonlinear
  inverse gravimetry
Authors: Max Kontak, Volker Michel

Inverse gravimetry: background material and multiscale mollifier approaches
Authors: Willi Freeden, M. Zuhair Nashed

Space-time GMsFEM for transport equations
Authors: Eric T. Chung, Yalchin Efendiev, Yanbo Li

Describing the singular behaviour of parabolic equations on cones in fractional
  Sobolev spaces
Authors: Stephan Dahlke, Cornelia Schneider

On variance component estimation with pseudo-observations
Author: E. Mysen

Algorithm of micro-seismic source localization based on asymptotic probability distribution of phase difference between two random stationary Gaussian processes
Authors: A. Varypaev, A. Kushnir

Simulation of settlement and bearing capacity of shallow foundations with soft
particle code (SPARC) and FE
Authors: Barbara Schneider-Muntau, Iman Bathaeian

Advanced computation of steady-state fluid flow in discrete Fracture-Matrix models: FEM–BEM and VEM–VEM fracture-block coupling
Authors: S. Berrone, A. Borio, C. Fidelibus, S. Pieraccini, S. Scialó, F. Vicini

A.5   Present State of GEM International Journal on Geomathematics (December 2018)

The above list contains a large number of distinguished scientists from mathematics
and geosciences, so that the scientific standard of the journal is extremely high.

Most acknowledged recent articles are the following publications:

Erik Mysen: On variance component estimation with pseudo-observations (Vol. 9,
Issue 2; 2018),

Yoshihito Kazashi: A fully discretised polynomial approximation on spherical shells
(Vol. 7, Issue 2; 2016),

Karl-Rudolf Koch: Monte Carlo methods (Vol. 9, Issue 1; 2018).

Being now a recognized journal the Editor-in-Chief decided to offer a new option to
the scientific community:

  "Topical   Collection",

where invited Guest Editors are responsible for a dedicated set of research articles.

The first "Topical Collections" to be published in GEM International Journal on
Geomathematics are as follows:

Numerical Methods for Processes in Fractured Porous Media.
Editors: I. Berre, L. Formaggia, A. Fumagalli, E. Keilegavlen, A. Scotti.

Mathematical Problems in Medical Imaging and Earth Sciences.
Editor: V. Michel.

Uncertainty Quantification in Subsurface Environments.
Editors: L. Tamellini, G. Porta, O. Ernst.

A.6   Future Obligations Concerning the GEM-Editorial Board

In response to the strong and positive feedback of the survey-like articles bridging mathematics and geosciences the decision was to encourage more submissions of this kind. To make this more visible from January 2019 on the founding Editor-in-Chief, Willi Freeden, decided to mainly focus on these articles and to appoint Volker Michel as successor Editor-in-Chief for the regular research articles.

This team of Editors-in-Chief will further develop the journal to an indispensable forum for research in geomathematics.

Appendix B
Handbook of Geomathematics

The *"Handbook of Geomathematics"* (W. Freeden, M.Z. Nashed, T. Sonar, eds.) consolidates the current knowledge by providing succinct summaries of concepts and theories, definitions of terms, biographical entries, organizational profiles, a guide to sources and information, and an overview to landscapes and contours of today's geomathematics. Contributions are written in an easy-to-understand and informative style for a general readership, typically from areas outside the particular research field.

The first edition of the "Handbook of Geomathematics" was published by Springer, Berlin, Heidelberg, in the year 2010. It broke new mathematical ground in dealing generically with fundamental problems and geoscientifically relevant "key technologies" as well as exploring the wide range of interactions and consequences for (wo)mankind. The list below shows the table of contents of the second (largely extended) edition published by Springer in 2015. It comprises the following scientific fields:

General Issues, Historical Background, and Future Perspectives,
Observational and Measurement Key Technologies,
Modeling of the System Earth,
Analytic, Algebraic, and Operator Theoretical Methods,
Statistical and Stochastic Methods,
Special Function Systems and Methods,
Computational and Numerical Methods,
Cartographic, Photogrammetric, Information Systems and Methods.

General Issues, Historical Background, and Future Perspectives

Geomathematics: Its Role, Its Aim, and Its Potential
Author: Willi Freeden

Navigation on Sea: Topics in the History of Geomathematics
Author: Thomas Sonar

© The Author(s), under exclusive license to Springer Nature Switzerland AG 2019
W. Freeden et al., *An Invitation to Geomathematics*, Lecture Notes in Geosystems
Mathematics and Computing, https://doi.org/10.1007/978-3-030-13054-1

Gravitational Viscoelastodynamics
Author: Detlef Wolf

Elastic and Viscoelastic Response of the Lithosphere to Surface Loading
Authors: V. Klemann, M. Thomas, H. Schuh

Multiscale Model Reduction with Generalized Multiscale Finite Element Methods
in Geomathematics
Authors: Yalchin Efendiev, Michael Presho

Efficient Modeling of Flow and Transport in Porous Media Using Multi-physics and
Multi-scale Approaches
Authors: Rainer Helmig, Bernd Flemisch, Markus Wolff, Benjamin Faigle

Convection Structures of Binary Fluid Mixtures in Porous Media
Authors: Matthias Augustin, Rudolf Umla, Manfred Lücke

Numerical Dynamo Simulations: From Basic Concepts to Realistic Models
Authors: Johannes Wicht, Stephan Stellmach, Helmut Harder

Mathematical Properties Relevant to Geomagnetic Field Modeling
Authors: Terence J. Sabaka, Gauthier Hulot, Nils Olsen

Multiscale Modeling of the Geomagnetic Field and Ionospheric Currents
Author: Christian Gerhards

Toroidal-Poloidal Decompositions of Electromagnetic Green's Functions in Geomagnetic Induction
Author: Jin Sun

Using B-Spline Expansions for Ionosphere Modeling
Authors: Michael Schmidt, Denise Dettmering, Florian Seitz

The Forward and Adjoint Methods of Global Electromagnetic Induction for
CHAMP Magnetic Data
Author: Zdeněk Martinec

Modern Techniques for Numerical Weather Prediction: A Picture Drawn from Kyrill
Authors: Nils Dorband, Martin Fengler, Andreas Gumann, Stefan Laps

Radio Occultation via Satellites
Authors: Christian Blick, Sarah Eberle

Asymptotic Models for Atmospheric Flows
Author: Rupert Klein

Stokes Problem, Layer Potentials and Regularizations, and Multiscale Applications
Authors: Carsten Mayer, Willi Freeden

On High Reynolds Number Aerodynamics: Separated Flows
Author: Mario Aigner

Turbulence Theory
Authors: Steffen Schön, Gaë Kermarrec

Forest Fire Spreading
Sarah Eberle, Willi Freeden, Ulrich Matthes

Phosphorus Cycles in Lakes and Rivers: Modeling, Analysis, and Simulation
Authors: Andreas Meister, Joachim Benz

Model-Based Visualization of Instationary Geo-Data with Application to Volcano
   Ash Data
Authors: Martin Baumann, Jochen Förstner, Vincent Heuveline, Jonas Kratzke,
   Sebastian Ritterbusch, Bernhard Vogel et al.

Modeling of Fluid Transport in Geothermal Research
Authors: Jörg Renner, Holger Steeb

Fractional Diffusion and Wave Propagation
Author: Yuri Luchko

Modeling Deep Geothermal Reservoirs: Recent Advances and Future Perspectives
Authors: Matthias Augustin, Mathias Bauer, Christian Blick, Sarah Eberle, Willi
   Freeden, Christian Gerhards, Maxim Ilyasov, René Kahnt, Mathias Klug, Sandra
   Möhringer, Thomas Neu, Isabel Michel née Ostermann, Allessandro Punzi

Analytic, Algebraic, and Operator Theoretical Methods

Noise Models for Ill-Posed Problems
Authors: Paul N. Eggermont, Vincent LaRiccia, M. Zuhair Nashed

Sparsity in Inverse Geophysical Problems
Authors: Markus Grasmair, Markus Haltmeier, Otmar Scherzer

Multiparameter Regularization in Downward Continuation of Satellite Data
Authors: Shuai Lu, Sergei V. Pereverzev

Evaluation of Parameter Choice Methods for Regularization of Ill-Posed Problems
   in Geomathematics
Authors: Frank Bauer, Martin Gutting, Mark A. Lukas

Quantitative Remote Sensing Inversion in Earth Science: Theory and Numerical
   Treatment
Author: Yanfei Wang

Correlation Modeling of the Gravity Field in Classical Geodesy
Authors: Christopher Jekeli

Inverse Resistivity Problems in Computational Geoscience
Authors: Alemdar Hasanov (Hasanoglu), Balgaisha Mukanova

Identification of Current Sources in 3D Electrostatics
Authors: Aron Sommer, Andreas Helfrich-Schkarbanenko, Vincent Heuveline

Transmission Tomography in Seismology
Author: Guust Nolet

Numerical Algorithms for Non-smooth Optimization Applicable to Seismic
    Recovery
Author: Ignace Loris

Strategies in Adjoint Tomography
Authors: Yang Luo, Ryan Modrak, Jeroen Tromp

Potential-Field Estimation Using Scalar and Vector Slepian Functions at Satellite
    Altitude
Authors: Alain Plattner, Frederik J. Simons

Multidimensional Seismic Compression by Hybrid Transform with Multiscale-
    Based Coding
Authors: Amir Z. Averbuch, Valery A. Zheludev, Dan D. Kosloff

Tomography: Problems and Multiscale Solutions
Author: Volker Michel

RFMP: An Iterative Best Basis Algorithm for Inverse Problems in the Geosciences
Author: Volker Michel

Material Behavior: Texture and Anisotropy
Authors: Ralf Hielscher, David Mainprice, Helmut Schaeben

Rayleigh Wave Dispersive Properties of a Vector Displacement as a Tool for P- and
    S-Wave Velocities Near Surface Profiling
Authors: Andrey Konkov, Andrey Lebedev, Sergey Manakov

Simulation of Land Management Effects on Soil N2O Emissions Using a Coupled
    Hydrology-Biogeochemistry Model on the Landscape Scale
Authors: Martin Wlotzka, Vincent Heuveline, Steffen Klatt, Edwin Haas, David
    Kraus, Klaus Butterbach-Bahl, Philipp Kraft, Lutz Breuer

Statistical and Stochastic Methods

An Introduction to Prediction Methods in Geostatistics
Authors: Ralf Korn, Alexandra Kochendörfer

Statistical Analysis of Climate Series
Author: Helmut Pruscha

Oblique Stochastic Boundary-Value Problem
Authors: Martin Grothaus, Thomas Raskop

Geodetic Deformation Analysis with Respect to an Extended Uncertainty Budget
Author: Hansjörg Kutterer

It's All About Statistics: Global Gravity Field Modeling from GOCE and Complementary Data
Author: Roland Pail

Mixed Integer Estimation and Validation for Next Generation GNSS
Author: Peter J. G. Teunissen

Mixed Integer Linear Models
Author: Peiliang Xu

Special Function Systems and Methods

Special Functions in Mathematical Geosciences: An Attempt at a Categorization
Authors: Willi Freeden, Michael Schreiner

Clifford Analysis and Harmonic Polynomials
Authors: Klaus Gürlebeck, Wolfgang Sprößig

Splines and Wavelets on Geophysically Relevant Manifolds
Author: Isaac Pesenson

Scalar and Vector Slepian Functions, Spherical Signal Estimation and Spectral Analysis
Authors: Frederik J. Simons, Alain Plattner

Dimension Reduction and Remote Sensing Using Modern Harmonic Analysis
Authors: John J. Benedetto, Wojciech Czaja

Computational and Numerical Methods

Radial Basis Function-Generated Finite Differences: A Mesh-Free Method for Computational Geosciences
Authors: Natasha Flyer, Grady B. Wright, Bengt Fornberg

Numerical Integration on the Sphere
Authors: Kerstin Hesse, Ian H. Sloan, Robert S. Womersley

Fast Spherical/Harmonic Spline Modeling
Author: Martin Gutting

Multiscale Approximation
Author: Stephan Dahlke

Sparse Solutions of Underdetermined Linear Systems
Authors: Inna Kozlov, Alexander Petukhov

Nonlinear Methods for Dimensionality Reduction
Authors: Charles K. Chui, Jianzhong Wang

Cartographic, Photogrammetric, Information Systems and Methods

Cartography
Author: Liqiu Meng

Theory of Map Projection: From Riemann Manifolds to Riemann Manifolds
Author: Erik W Grafarend

Modeling Uncertainty of Complex Earth Systems in Metric Space
Authors: Jef Caers, Kwangwon Park, Céline Scheidt

Geometrical Reference Systems
Authors: Manuela Seitz, Detlef Angermann, Michael Gerstl, Mathis Bloßfeld, Laura
    Sánchez, Florian Seitz

Analysis of Data from Multi-satellite Geospace Missions
Author: Joachim Vogt

Geodetic World Height System Unification
Author: Michael Sideris

Mathematical Foundations of Photogrammetry
Author: Konrad Schindler

Potential Methods and Geoinformation Systems
Author: Hans-Jürgen Götze

Geoinformatics
Author: Monika Sester

Appendix C
Published Books and Edited Handbooks by Willi Freeden

W. Freeden (Ed.). Progress in Geodetic Science (at GW 98). Shaker, Aachen, 1998

W. Freeden, T. Gervens, and M. Schreiner. Constructive Approximation on the Sphere (With Application to Geomathematics). Oxford Sciences Publication. Clarendon Press, Oxford University, 1998

W. Freeden. Multiscale Modelling of Spaceborne Geodata. Teubner Verlag, Stuttgart, Leipzig, 1999

W. Freeden, V. Michel. Multiscale Potential Theory (With Applications to Geoscience). Birkhäuser Verlag, Boston, Basel, Berlin, 2004

W. Freeden, M. Schreiner. Spherical Functions of Mathematical Geosciences. A Scalar, Vectorial, and Tensorial Setup. Springer, Heidelberg, 2009

W. Freeden, Z. Nashed, T. Sonar (eds.). Handbook of Geomathematics, 1st ed., Springer, Heidelberg 2010

W. Freeden, Metaharmonic Lattice Point Theory. CRC Press, Taylor & Francis Group, Boca Raton, 2011

W. Freeden, C. Gerhards. Geomathematically Oriented Potential Theory. CRC Press, Taylor & Francis Group, Boca Raton, 2013

W. Freeden, M. Gutting. Special Functions of Mathematical (Geo-)Physics. Birkhäuser, Basel, 2013

M. Bauer, W. Freeden, H. Jacobi, T. Neu (Herausgeber). Handbuch Tiefe Geothermie, Springer Spektrum, Heidelberg, 2014

W. Freeden, Z. Nashed, T. Sonar (Eds.). Handbook of Geomathematics. 2nd ed., Springer, Heidelberg, 2015

W. Freeden, R. Rummel (Herausgeber). Photogrammetrie und Fernerkundung (C. Heipke, ed.), Springer Spektrum, Heidelberg, 2017

W. Freeden, R. Rummel (Herausgeber). Erdmessung und Satellitengeodäsie. (Rummel, ed.), Springer Spektrum, Heidelberg, 2017

© The Author(s), under exclusive license to Springer Nature Switzerland AG 2019
W. Freeden et al., *An Invitation to Geomathematics*, Lecture Notes in Geosystems Mathematics and Computing, https://doi.org/10.1007/978-3-030-13054-1

W. Freeden, M. Gutting. Integration and Cubature Methods - A Geomathematically Oriented Course, CRC Press, Taylor and Francis Group, Boca Raton, 2017

W. Freeden, R. Rummel (Herausgeber). Ingenieurgeodäsie (W. Schwarz, ed.), Springer Spektrum, Heidelberg, 2018

M. Bauer, W. Freeden, H. Jacobi, T. Neu (Herausgeber). Handbuch Oberflächennahe Geothermie, Springer Spektrum, Heidelberg, 2018

W. Freeden, M.Z. Nashed (eds.). Handbook of Mathematical Geodesy, Geosystems Mathematics, Birkhäuser, Springer International Publishing AG, Basel, 2018

W. Freeden, M.Z. Nashed, M. Schreiner. Spherical Sampling, Geosystems Mathematics, Birkhäuser, Springer International Publishing AG, Basel, 2018

W. Freeden, R. Rummel (Herausgeber). Mathematische Geodäsie/Mathematical Geodesy (W. Freeden, ed.), Springer Spektrum, Heidelberg, 2019

Appendix D
PhD-Dissertations in Geomathematics Supervised by Willi Freeden

R. Reuter (1982): "Über Integralformeln der Einheitssphäre und harmonische Splinefunktionen"
Korreferent: F. Reuter (Aachen)

H. Schaffeld (1988): "Finite-Elemente-Methoden und ihre Anwendung zur Erstellung von Digitalen Geländemodellen"
Korreferenten: H. Esser (Aachen), B. Witte (Aachen)

M. Schreiner (1994): "Tensor Spherical Harmonics and Their Application in Satellite Gradiometry"
Korreferent: R. Rummmel (Delft)

J. Cui (1995): "Finite Pointset Methods on the Sphere and Their Application in Physical Geodesy"
Korreferent: H. Sünkel (Graz)

U. Windheuser (1995): "Sphärische Wavelets: Theorie und Anwendungen in der Physikalischen Geodäsie"
Korreferent: P. Maaß (Potsdam)

M. Tücks (1996): "Navier-Splines und ihre Anwendung in der Deformationanalyse"
Korreferenten: E. Groten (Darmstadt), S.L. Svensson (Lund)

F. Schneider (1997): "Inverse Problems in Satellite Geodesy and Their Approximate Solution by Splines and Wavelets"
Korreferent: E. Schock (Kaiserslautern)

V. Michel (1999): "A Multiscale Method for the Gravimetry Problem: Theoretical and Numerical Aspects of Harmonic and Anharmonic Modelling"
Korreferenten: E. Groten (Darmstadt), E. Schock (Kaiserslautern)

© The Author(s), under exclusive license to Springer Nature Switzerland AG 2019   117
W. Freeden et al., *An Invitation to Geomathematics*, Lecture Notes in Geosystems
Mathematics and Computing, https://doi.org/10.1007/978-3-030-13054-1

M. Bayer (1999): "Geomagnetic Field Modelling From Satellite Data by First and Second Generation Wavelets"
Korreferenten: H. Lühr (Potsdam), S.L. Svensson (Lund)

S. Beth (2000): "Multiscale Approximation by Vector Radial Basis Functions on the Sphere"
Korreferenten: J. Mason (Shrivenham), B. Witte (Bonn)

O. Glockner (2001): "On Numerical Aspects of Gravitational Field Modelling from SST and SGG by Harmonic Splines and Wavelets (With Application to CHAMP Data)"
Korreferenten: J. Kusche (Delft), H. Sünkel (Graz)

H. Nutz (2001): "A Unified Setup of Gravitational Field Observables"
Korreferenten: J. Prestin (Lübeck), R. Rummel (München)

R. Litzenberger (2001): "Pyramid Schemes for Harmonic Wavelets in Boundary–Value Problems"
Korreferent: E. Schock (Kaiserslautern)

T. Maier (2002): "Multiscale Geomagnetic Field Modelling From Satellite Data: Theoretical Aspects and Numerical Applications"
Korreferent: N. Olsen (Kopenhagen)

K. Hesse (2003): "Domain Decomposition Methods in Multiscale Geopotential Determination from SST and SGG"
Korreferenten: E. Groten (Darmstadt), I. Sloan (Sydney)

M.K. Abeyratne (2003): "Cauchy-Navier Wavelet Solvers and Their Application in Deformation Analysis"
Korreferent: E. Groten (Darmstadt)

C. Mayer (2003): "Wavelet Modelling of Ionospheric Currents and Induced Magnetic Fields From Satellite Data"
Korreferent: H. Lühr (Potsdam)

F. Bauer (2004): "An Alternative Approach to the Oblique Derivative Problem in Potential Theory"
Korreferent: S. Pereverzev (Linz)

M. J. Fengler (2005): "Vector Spherical Harmonic and Vector Wavelet Based Non-Linear Galerkin Schemes for Solving the Incompressible Navier-Stokes Equation on the Sphere"
Korreferent: T. Sonar (Braunschweig)

S. Gramsch (2006): "Integralformeln und Wavelets auf regulären Gebieten der Sphäre"
Korreferent: M. Schreiner (Buchs)

A. Luther (2007): "Vector Field Approximation on Regular Surfaces in Terms of Outer Harmonic Representations"
Korreferent: G. Schüler (Trier)

M. Gutting (2007): "Fast Multipole Methods for Oblique Derivative Problems"
Korreferent: O. Steinbach (Graz)

A. Moghiseh (2007): "Fast Wavelet Transform by Biorthogonal Locally Supported Radial Bases Functions on Fixed Spherical Grids"
Korreferent: M. Schreiner (Buchs)

O. Schulte (2009): "Euler Summation Oriented Spline Interpolation"
Korreferent: E.W. Grafarend (Stuttgart)

T. Fehlinger (2009): "Multiscale Formulations for the Disturbing Potential and the Deflections of the Vertical in Locally Reflected Physical Geodesy"
Korreferent: P. Holota (Prag)

K. Wolf (2009): "Multiscale Modeling of Classical Boundary Value Problems in Physical Geodesy by Locally Supported Wavelets"
Korreferent: R. Rummel (München)

A. Kohlhaas (2010): "Multiscale Methods on Regular Surfaces and their Application to Physical Geodesy"
Korreferent: E. Groten (Darmstadt)

C. Gerhards (2011): "Spherical Multiscale Methods in Terms of Locally Supported Wavelets: Theory and Application to Geomagnetic Modeling"
Korreferent: N. Olsen (Kopenhagen)

I. Ostermann (2011): "Modeling Heat Transport in Deep Geothermal Systems by Radial Basis Functions"
Korreferent: R. Helmig (Stuttgart)

M. Ilyasov (2011): "A Tree Algorithm for Helmholtz Potential Wavelets on Non-smooth Surfaces: Theoretical Background and Application to Seismic Data Processing"
Korreferent: M. Popov (Petersburg)

E. Kotevska (2011): "Real Earth Oriented Gravitational Potential Determination"
Korreferent: H. Schaeben (Freiberg)

S. Möhringer (2014): "Decorrelation of Gravimetric Data"
Korreferent: J. Kusche (Bonn)

M. Klug (2014): "Integral Formulas and Discrepancy Estimates Using the Fundamental Solution to the Beltrami Operator on Regular Surfaces"
Korreferent: E.W. Grafarend (Stuttgart)

S. Eberle (2014): "Forest Fire Determination: Theory and Numerical Aspects"
Korreferent: L. Ferragut Canals (Salamanca)

M. Augustin (2015): "A Method of Fundamental Solutions in Poroelasticity to Model the Stress Field in Geothermal Reservoirs"
Korreferent: T. Sonar (Braunschweig)

C. Blick (2015): "Multiscale Potential Methods in Geothermal Research: Decorrelation Reflected Post-Processing and Locally Based Inversion"
Korreferent: F. J. Simons (Princeton)

References

1. Achenbach, J.D.: Wave Propagation in Elastic Solids. North Holland Publishing Company, New York (1973)
2. Agterberg, F.P: Geomathematics. Elsevier Scientific Publishing Company, Rotterdam (1974)
3. Aronszajn, N.: Theory of reproducing kernels. Trans. Am. Math. Soc. **68**, 337–404 (1950)
4. Backus, G.E., Gilbert, F.: Numerical applications of a formalism for geophysical inverse problems. Geophys. J. R. Astron. Soc. **13**, 247–276 (1967)
5. Backus, G.E., Gilbert, F.: The resolving power of gross Earth data. Geophys. J. R. Astron. Soc. **16**, 169–205 (1968)
6. Backus, G.E., Gilbert, F.: Uniqueness in the inversion of inaccurate gross Earth data. Philos. Trans. R. Soc. Lond. A **266**, 123–192 (1970)
7. Baysal, E., Kosloff, D.D., Sherwood, J.W.C.: A two-way nonreflecting wave equation. Geophysics **49**, 132–141 (1984)
8. Beylkin, G., Monzón, L.: On approximation of functions by exponential sums. Appl. Comput. Harmon. Anal. **19**, 17–48 (2005)
9. Beylkin, G., Monzón, L.: Approximation of functions by exponential sums revisited. Appl. Comput. Harmon. Anal. **28**, 131–149 (2010)
10. Blick, C., Freeden, W., Nutz, H.: Gravimetry and Exploration. In: Freeden, W., Nashed, M.Z. (eds.) Handbook of Mathematical Geodesy. Geosystems Mathematics, pp. 687–752. Birkhäuser/Springer, Basel/New-York/Heidelberg (2018)
11. Burschäpers, H.C.: Local modeling of gravitational data. Master Thesis, University of Kaiserslautern, Mathematics Department, Geomathematics Group (2013)
12. Cheng, H., Greengard, L., Rokhlin, V.: A fast adaptive multipole algorithm in three dimensions. J. Comput. Phys. **155**, 468–498 (1999)
13. Claerbout, J.: Basic Earth Imaging. Standford University, Standford (2009)
14. Cohen, L.: Time-Frequency Analysis. Prentice Hall, Englewood Clifffs (1995)
15. Daubechies, I.: Ten Lectures on Wavelets. In: CBMS-NSF Regional Conference Series in Applied Mathematics, vol. 61. SIAM, Philadelphia (1992)
16. Davis, P.J.: Interpolation and Approximation. Blaisdell, New York (1963)
17. de Laplace, P.S.: Theorie des attractions des sphéroides et de la figure des planètes. Mèm. de l'Acad., Paris (1785)
18. Eggermont, P.N., LaRiccia, V., Nashed, M.Z.: Noise Models for Ill-Posed Problems. In: Freeden, W., Nashed, M.Z., Sonar, T. (eds.) Handbook of Geomathematics, vol. 2, 2nd edn., pp. 1633–1658. Springer, New York (2015)
19. Engl, H.W., Hanke, M., Neubauer, A.: Regularization of Inverse Problems. Kluwer, Dordrecht (1996)

20. Evans, L.D.: Partial Differential Equation, Third Printing. American Mathematical Society, Providence (2002)
21. Freeden, W.: On spherical spline interpolation and approximation. Math. Methods Appl. Sci. **3**, 551–575 (1981)
22. Freeden, W.: On approximation by harmonic splines. Manuscr. Geodaet. **6**, 193–244 (1981)
23. Freeden, W.: A spline interpolation method for solving boundary value problems of potential theory from discretely given data. Numer. Methods Partial Differ. Equ. **3**, 375–398 (1987)
24. Freeden, W.: The Uncertainty Principle and Its Role in Physical Geodesy. In: W. Freeden (ed.) Progress in Geodetic Science at GW 98, pp. 225–236. Shaker, Aachen (1998)
25. Freeden, W.: Multiscale Modelling of Spaceborne Geodata. B.G. Teubner, Stuttgart (1999)
26. Freeden, W.: Geomathematik, was ist das überhaupt? Jahresb. Deutsch. Mathem. Vereinigung (DMV) **111**, 125–152 (2009)
27. Freeden, W.: Geomathematics: Its Role, Its Aim, and Its Potential. In: Freeden, W., Nashed, M.Z., Sonar, T. (eds.) Handbook of Geomathematics, vol. 1, 2nd edn, pp. 3–78. Springer, Heidelberg (2015)
28. Freeden, W.: Handbook of Mathematical Geodesy: Introduction. In: Freeden, W., Nashed, M.Z. (eds.) Handbook of Mathematical Geodesy. Geosystems Mathematics, pp. 753–820. Springer, Basel (2018)
29. Freeden, W., Blick, C.: Signal decorrelation by means of multiscale methods. World Min. **65**, 1–15 (2013)
30. Freeden, W., Gerhards, C.: Geomathematically Oriented Potential Theory. Chapman and Hall/CRC Press, Boca Raton/London (2013)
31. Freeden, W., Gutting, M.: Special Functions of Mathematical (Geo)Physics. Birkhäuser, Basel (2013)
32. Freeden, W., Gutting, M.: Integration and Cubature Methods - A Geomathematically Oriented Course. Chapman and Hall/CRC Press, Boca Raton/London/New York (2018)
33. Freeden, W., Maier, T.: On multiscale denoising of spherical functions: Basic theory and numerical aspects. Electron. Trans. Numer. Anal. **14**, 40–62 (2002)
34. Freeden, W., Maier, T.: Spectral and multiscale signal-to-noise thresholding of spherical vector fields. Comput. Geosci. **7**, 215–250 (2003)
35. Freeden, W., Michel, V.: Multiscale Potential Theory (With Applications to Geoscience). Birkhäuser, Boston (2004)
36. Freeden, W., Nashed, M.Z.: Multivariate Hardy-type lattice point summation and Shannon-type sampling. GEM Int. J. Geomath. **6**, 163–249 (2015)
37. Freeden, W., Nashed, M.Z.: Operator-theoretic and regularization approaches to ill-posed problems. GEM Int. J. Geomath. **9**, 1–115 (2017)
38. Freeden, W., Nashed, M.Z.: From Gaussian Circle Problem to Multivariate Shannon Sampling. In: Nashed, M.Z., Li, X. (eds.) Contemporary Mathematics and Its Applications. Frontiers in Orthogonal Polynomials and q-Series, vol. 1, pp. 213–238 (2017)
39. Freeden, W., Nashed, M.Z.: Inverse gravimetry: Background material and multiscale mollifier approaches. GEM Int. J. Geomath. **9**, 199–264 (2018)
40. Freeden, W., Nashed, M.Z.: Ill-Posed Problems: Operator Methodologies of Resolution and Regularization. In: Freeden, W., Nashed, M.Z. (eds.) Handbook of Mathematical Geodesy. Geosystems Mathematics, pp. 201–314. Springer, Basel (2018)
41. Freeden, W., Nashed, M.Z.: Gravimetry As an Ill-Posed Problem in Mathematical Geodesy. In: Freeden, W., Nashed, M.Z. (eds.) Handbook of Mathematical Geodesy. Geosystems Mathematics, pp. 641–686. Springer, Basel (2018)
42. Freeden, W., Nutz, H.: Mathematische Methoden. In: Bauer, M., Freeden, W., Jacobi, H., Neu, T. (Herausgeber) Handbuch Tiefe Geothermie. Springer, Heidelberg (2014)
43. Freeden, W., Nutz, H.: Mathematik als Schlüsseltechnologie zum Verständnis des Systems "Tiefe Geothermie". Jahresber. Deutsch. Math. Vereinigung (DMV) **117**, 45–84 (2015)
44. Freeden, W., Nutz, H.: Geodetic Observables and Their Mathematical Treatment in Multiscale Framework. In: Freeden, W., Nashed, M.Z. (eds.) Handbook of Mathematical Geodesy. Geosystems Mathematics, pp. 315–458. Springer, Basel (2018)

45. Freeden, W., Sansó, F.: Geodesy and Mathematics: Interactions, Acquisitions, and Open Problems. In: International Association of Geodesy Symposia (IAGS), IX Hotine-Marussi Symposium Rome. Springer, Heidelberg (submitted, 2019). Preprint (2019)
46. Freeden, W., Schreiner, M.: Spherical Functions of Mathematical Geosciences – A Scalar, Vectorial, and Tensorial Setup. Springer, Heidelberg (2009)
47. Freeden, W., Schreiner, M.: Mathematical Geodesy: Its Role, Its Potential and Its Perspective. In: Freeden, W., Rummel, R. (eds.) Handbuch der Geodäsie. Springer Reference Naturwissenschaften. Springer, Cham (2019). https://doi.org/10.1007/978-3-662-46900-2_91_1
48. Freeden, W., Witte, B.: A combined (spline-) interpolation and smoothing method for the determination of the gravitational potential from heterogeneous data. Bull. Géod. **56**, 53–62 (1982)
49. Freeden, W., Gervens, T., Schreiner, M.: Constructive Approximation on the Sphere (With Applications to Geomathematics). Oxford Science Publications, Clarendon/Oxford (1998)
50. Freeden, W., Michel, V., Simons, F.J.: Spherical Harmonics Based Special Function Systems and Constructive Approximation Methods. In: Freeden, W., Nashed, M.Z. (eds.) Handbook of Mathematical Geodesy. Geosystems Mathematics, pp. 753–820. Springer, Basel (2018)
51. Freeden, W., Nashed, M.Z., Schreiner, M.: Spherical Sampling. Geosystems Mathematics. Springer, Basel (2018)
52. Freeden, W., Sonar, T., Witte, B.: Gauss as Scientific Mediator Between Mathematics and Geodesy from the Past to the Present. In: Freeden, W., Nashed, M.Z. (eds.) Handbook of Mathematical Geodesy, pp. 1–164. Geosystems Mathematics. Springer, Basel (2018)
53. Gauss, C.F.: Allgemeine Theorie des Erdmagnetismus. Resultate aus den Beobachtungen des magnetischen Vereins (1838)
54. Grafarend, E.W.: Six Lectures on Geodesy and Global Geodynamics. In: Moritz, H., Sünkel, H. (eds.) Proceedings of the Third International Summer School in the Mountains, pp. 531–685 (1982)
55. Grafarend, E.W., Klapp, M., Martinec, Z.: Spacetime Modelling of the Earth's Gravity Field by Ellipsoidal Harmonics. In: Freeden, W., Nashed, M.Z., Sonar, T. (eds.) Handbook of Geomathematics, vol. 1, 1st. edn., pp. 159–253. Springer, Heidelberg (2010)
56. Greengard, L., Rokhlin, V.: A new version of the fast multipole method for the Laplace equation in three dimensions. Acta Numer. **6**, 229–269 (1997)
57. Groten, E.: Geodesy and the Earth's Gravity Field I + II. Dümmler, Bonn (1979)
58. Gutting, M.: Fast multipole methods for oblique derivative problems. Ph.D. thesis, University of Kaiserslautern, Mathematics Department, Geomathematics Group (2007)
59. Gutting, M.: Fast multipole accelerated solution of the oblique derivative boundary value problem. GEM Int. J. Geomath. **3**, 223–252 (2012)
60. Gutting, M.: Fast Spherical/Harmonic Spline Modeling. In: Freeden, W., Nashed, Z., Sonar, T. (eds.) Handbook of Geomathematics, vol. 3, 2nd edn., pp. 2711–2746. Springer, New York (2015)
61. Hackbusch, W.: Entwicklungen nach Exponentialsummen. Technical Report. Max-Planck-Institut für Mahematik in den Naturwissenschaften, Leipzig (2010)
62. Hackbusch, W., Khoromoskij, B.N., Klaus, A.: Approximation of functions by exponential sums based on the Newton-type optimisation. Technical Report, Max-Planck-Institut für Mathematik in den Naturwissenschaften, Leipzig (2005)
63. Hadamard, J.: Sur les problèmes aux dérivées partielles et leur signification physique. Princet. Univ. Bull. **13**, 49–52 (1902)
64. Heiskanen, W.A., Moritz, H.: Physical Geodesy. Freeman, San Francisco (1967)
65. Helmert, F.: Die Mathematischen und Physikalischen Theorien der Höheren Geodäsie, I, II. B.G. Teubner, Leipzig (1884)
66. Hille, E.: Introduction to the general theory of reproducing kernels. Rocky Mountain J. Math. **2**, 321–368 (1972)
67. Hofmann-Wellenhof, B., Moritz, H.: Physical Geodesy. Springer, Wien (2005)
68. Ilyasov, M.: A tree algorithm for Helmholtz potential wavelets on non-smooth surfaces: theoretical background and application to seismic data processing. Ph.D. thesis, Geomathematics Group, University of Kaiserslautern (2011)

69. Jakobs, F., Meyer, H.: Geophysik – Signale aus der Erde. Teubner, Leipzig (1992)
70. Jansen, M., Oonincx, P.: Second Generation Wavelets and Applications. Springer, Berlin (2005)
71. Klein, F.: Elementarmathematik III. Die Grundlagen der Mathematischen Wissenschaften, Band 16. Springer, Berlin (1928)
72. Legendre, A.M.: Recherches sur l'attraction des sphèroides homogènes. Mèm. math. phys. près. à l'Acad. Aci. par. divers savantes **10**, 411–434 (1785)
73. Listing, J.B.: Über unsere jetzige Kenntnis der Gestalt und Größe der Erde. Dietrichsche Verlagsbuchhandlung, Göttingen (1873)
74. Louis, A.K.: Inverse und schlecht gestellte Probleme. Teubner, Stuttgart (1989)
75. Mallat, S.: Applied Mathematics Meets Signal Proeessing. In: Proceedings of the International Congress of Mathematicians, Berlin 1998, vol. I, pp. 319–338. Documenta Mathematica, Bielefeld (1998)
76. Marks, D.L.: A family of approximations spanning the Born and Rytov scattering series. Opt. Express **14**, 8837–8848 (2013)
77. Martin, G.S., Marfurt, K.J., Larsen, S.: Marmousi-2: An updated model for the investigation of AVO in structurally complex areas. In: Proceedings, SEG Annual Meeting, Salt Lake City (2002)
78. Martin, G.S., Wiley, R., Marfurt, K.J.: Marmousi2: An elastic upgrade for marmousi. Lead. Edge **25**, 156–166 (2006)
79. Marussi, A.: Intrinsic Geodesy. Springer, Berlin (1985)
80. Meissl, P.A.: A study of covariance functions related to the Earth's disturbing potential. Department of Geodetic Science, Report No. 151, The Ohio State University, Columbus, OH (1971)
81. Meissl, P.A.: Hilbert spaces and their applications to geodetic least squares problems. Boll. Geod. Sci. Aff. **1**, 181–210 (1976)
82. Michel, V.: A multiscale method for the gravimetry problem: theoretical and numerical aspects of harmonic and anharmonic modelling. Ph.D. thesis, University of Kaiserslautern, Mathematics Department, Geomathematics Group, Shaker, Aachen (1999)
83. Michel, V.: Lectures on Constructive Approximation. Applied and Numerical Harmonic Analysis. Birkhäuser, New York (2013)
84. Michel, V., Fokas, A.S.: A unified approach to various techniques for the non-uniqueness of the inverse gravimetric problem and wavelet-based methods. Inverse Prob. **24** (2008). https://doi.org/10.1088/0266-5611/24/4/045019
85. Möhringer, S.: Decorrelation of gravimetric data. Ph.D. thesis, University of Kaiserslautern, Mathematics Department, Geomathematics Group (2014)
86. Moritz, H.: Advanced Physical Geodesy. Herbert Wichmann Verlag/Abacus Press, Karlsruhe/Tunbridge (1980)
87. Moritz, H.: Geodesy and Mathematics. Zeszyty Naukowe Akademii Górniczo-Hutniczej IM Stanislawa Staszica, No. 780. Geodezja, vol. 63, pp. 38–43, Kraków (1981)
88. Moritz, H.: The Figure of the Earth. Theoretical Geodesy of the Earth's Interior. Wichmann Verlag, Karlsruhe (1990)
89. Müller, C., Aspects of Differential Equations in Mathematical Physics. In: Langer, R.E. (ed.) Partial Differential Equations and Continuum Mechanics, pp. 3–8. The University of Wisconsin Press, Madison (1961)
90. Müller, C.: Foundations of the Mathematical Theory of Electromagnetic Waves. Springer, Berlin (1969)
91. Nashed, M.Z.: New Applications of Generalized Inverses in System and Control Theory. In: Thomas, J.B. (ed.) Proceedings of 1980 Conference on Information Sciences and Systems, Princeton, pp. 353–358 (1980)
92. Nashed, M.Z.: Operator-theoretic and computational approaches to ill-posed problems with applications to antenna theory. IEEE Trans. Antennas Propag. **29**, 220–231 (1981)
93. Nashed, M.Z.: A New Approach to Classification and Regularization of Ill-Posed Operator Equations. In: Engl, H., Groetsch, C.W. (eds.) Inverse and Ill-Posed Problems, Band 4. Notes and Reports in Mathematics and Science and Engineering. Academic Press, Boston (1987)

94. Nashed, M.Z.: Inverse Problems, Moment Problems and Signal Processing: Un Menage a Trois. In: Siddiqi, A.H., Singh, R.C., Manchanda, P. (eds.) Mathematics in Science and Technology, pp. 1–19. World Scientific, New Jersey (2010)

95. Nashed, Z.M., Sun, Q.: Function Spaces for Sampling Expansions. In: Shen, A.I., Zayed, A. (eds.) Multiscale Signal Analysis and Modeling, pp. 81–104. Springer, New York (2013)

96. Nashed, M.Z., Wahba, G.: Generalized inverses in reproducing kernel spaces: an approach to regularization of linear operator equations. SIAM J. Math. Anal. **5**, 974–987 (1974)

97. Nashed, M.Z., Wahba, G.: Regularization and approximation of liner operator equations in reproducing kernel spaces. Bull. Am. Math. Soc. **80**, 1213–1218 (1974)

98. Nashed, M.Z., Walter, G.G.: General sampling theorems for functions in reproducing kernel Hilbert spaces. Math. Control Signals Syst. **4**, 363–390 (1991)

99. Nashed, M.Z., Walter, G.G.: Reproducing kernel Hilbert space from sampling expansions. Contemp. Math. **190**, 221–226 (1995)

100. Nolet, G.: Seismic Tomography: Imaging the Interior of the Earth and Sun. Cambridge University Press, Cambridge (2008)

101. Nutz, H.: A unified setup of gravitational field observables. Ph.D. thesis, University of Kaiserslautern, Geomathematics Group, Shaker, Aachen (2002)

102. Pavlis, N.K., Holmes, S.A., Kenyon, S.C., Factor, J.K.: An Earth Gravitational Model to Degree 2160: EGM2008. General Assembly of the European Geosciences Union, Vienna (2018)

103. Popov, M.M., Semtchenok, N.M., Popov, P. M., Verdel, A.R.: Gaussian beam migration of multi-valued zero-offset data. In: Proceedings, International Conference, Days on Diffraction, St Petersburg, Russia, pp. 225–234 (2006)

104. Popov, M.M., Semtchenok, N.M., Popov, P.M., Verdel, A.R.L.: Reverse time migration with gaussian beams and velocity analysis applications. In: Extended Abstracts, 70th EAGE Conference & Exhibitions, Rome, F048 (2008)

105. Potts, D., Steidl, G., Tasche, M.: Fast Fourier transforms for nonequispaced data: a tutorial, modern sampling theory. Appl. Numer. Harmon. Anal. (ACHA) **22**, 247–270 (2001)

106. Rummel, R.: Geodesy. In: Encyclopedia of Earth System Science, vol. 2, pp. 253–262. Academic, New York (1992)

107. Rummel, R.: Spherical Spectral Properties of the Earth's Gravitational Potential and Its First and Second Derivatives. In: Lecture Notes in Earth Science, vol. 65, pp. 359–404. Springer, Berlin (1997)

108. Rummel, R., van Gelderen, M.: Meissl scheme – spectral characteristics of physical geodesy. Manuscr. Geod. **20**, 379–385 (1995)

109. Saitoh, S.: Theory of Reproducing Kernels and Its Applications. Longman, New York (1988)

110. Shure, L., Parker, R.L., Backus, G.E.: Harmonic splines for geomagnetic modelling. Phys. Earth Planet. Inter. **28**, 215–229 (1982)

111. Skudrzyk, E.: The Foundations of Acoustics. Springer, Heidelberg (1972)

112. Snieder, R.: The Perturbation Method in Elastic Wave Scattering and Inverse Scattering in Pure and Applied Science. General Theory of Elastic Waves, pp. 528–542. Academic, San Diego (2002)

113. Sonar, T.: 3000 Jahre Analysis. Springer, Berlin (2011)

114. Stokes, G.G.: On the variation of gravity at the surface of the Earth. Trans. Camb. Philos. Soc. **148**, 672–712 (1849)

115. Svensson, S.L.: Pseudodifferential operators: a new approach to the boundary value problems of physical geodesy. Manuscr. Geod. **8**, 1–40 (1983)

116. Symes, W.W.: The Rice Inversion Project, Department of Computational and Applied Mathematics, Rice University, Houston, Texas, USA. http://www.trip.caam.rice.edu/downloads/downloads.html. Accessed 12 Sept 2016

117. Torge, W.: Gravimetry. de Gruyter, Berlin (1989)

118. Torge, W.: Geodesy. de Gruyter, Berlin (1991)

119. Wahba, G.: Spline interpolation and smoothing on the sphere. SIAM J. Sci. Stat. Comput. **2**, 5–16. Also: Errata SIAM J. Sci. Stat. Comput. **3**, 385–386 (1981)

120. Wahba, G.: Spline Models for Observational Data. In: CBMS-NSF Regional Conference Series in Applied Mathematics, vol. 59. SIAM, Philadelphia (1990)
121. Weck, N.: Zwei inverse Probleme in der Potentialtheorie. Mitt. Inst. Theor. Geodäsie, Universität Bonn **4**, 27–36 (1972)
122. Xia, X.G., Nashed, M.Z.: The Backus-Gilbert method for signals in reproducing Hilbert spaces and wavelet subspaces. Inverse Prob. **10**, 785–804 (1994)
123. Yilmas, O.: Seismic Data Analysis: Processing, Inversion and Interpretation of Seismic Data. Society of Exploration Geophysicists, Tulsa (1987)

Index

Printed in the United States
by Baker & Taylor Publisher Services